PAPUA NEW GUINEA

Making a Living

Book 1

Pam Norman
Stephen Potek
Eron Hagunama
Joe Deruage

Oxford University Press is a department of the University of Oxford. It furthers the University's objective of excellence in research, scholarship, and education by publishing worldwide. Oxford is a registered trademark of Oxford University Press in the UK and in certain other countries.

Published in Australia by
Oxford University Press
Level 8, 737 Bourke Street, Docklands, Victoria 3008, Australia

First published 2005
Reprinted 2006, 2007, 2008, 2009, 2012, 2013 (third), 2014 (twice), 2016, 2017 (twice), 2018, 2022 (twice), 2023, 2025

ISBN 978 0 19 555116 7

Typeset by Lynn Twelftree
Printed in China by Golden Cup Printing Co. Ltd.

Oxford University Press Australia & New Zealand is committed to sourcing paper responsibly.

Acknowledgement
This book was developed with the support of the Australian Government through the Curriculum Reform Implementation Project.

Contents

(continues)

Contents

Secretary's Message

Making a Living is a new subject area that aims to empower students with the essential life skills and knowledge to participate actively in society, to be self-reliant and to live sustain ably now and in the future.

It combines the subjects of Commerce, Agriculture and Technology to promote useful and relevant life skills. It has been allocated 6 hours (360 minutes) per week to give teachers enough time to teach this very important subject.

Students will manage resources, promote better living and participate actively in community development. This subject depends on initiatives, dedication and commitment from teachers, students and the community. Its success will be evident through practical projects in the school and local community.

Making a Living is an important part of the reform curriculum that emphasises the need to teach our students to be enterprising and improve standards of living as called for in the Matane report: *A Philosophy of Education for Papua New Guinea.*

Students will investigate, create and design innovative ways of using available resources. They will make objects, cost them and evaluate the quality of their products and processes. This subject is relevant and useful for students leaving formal schooling after Grade 8 as well as those who will pursue studies in higher learning institutions.

Student Books 1 and 2, and the accompanying Teacher Resource Book, have been prepared to provide direct support for the achievement of syllabus outcomes. The content is consistent with, and promotes the teaching and learning models of, the syllabus and the related Department of Education Teachers Guide. These resources encompass a range of material including student activities, advice to teachers, assessment ideas, homework activities, project activities, reference information and a glossary of terms. They are intended to facilitate flexible learning activity for teachers and students, and may be modified and amended to suit local circumstances.

I commend and approve these materials for use in all Upper Primary schools throughout Papua New Guinea.

Peter M. Baki

Peter M. Baki CBE
Secretary for Education

Managing Resources

Chapter Summary

In this chapter you will have an opportunity to:

- ✔ Identify land and water resources and investigate how best to manage these resources in a way that benefits people and the natural environment.
- ✔ Identify elements of the local environment (natural and human) and identify positive actions to preserve this environment.
- ✔ Learn principles and techniques of managing land to increase the quality and quantity of food production for consumption and for sale.
- ✔ Learn principles and techniques of managing livestock production to increase consumption and for sale.

Syllabus references

Strand: Managing Resources

Substrand 1: Land and Water Management

Outcome: 6.1.1 Investigate the importance of land and water resources and apply ways of managing these to meet basic human needs.

Substrand 2: Environment

Outcome: 6.1.2 Investigate the consequences when an element of the local environment is affected and apply environmentally friendly actions to care for it.

Substrand 3: Crops and Animal Management

Outcome: 6.1.3 Share an understanding of the economic, cultural and nutritional value of crops and animals and compare the benefits of traditional and commercial crops and animal management.

Land and Water Resources SS

A resource is anything we use. Banana is a food resource for many Papua New Guineans; kunai grass is a resource traditionally used for building houses and mangrove tree is a resource for building over the sea.

Papua New Guinea has traditionally had subsistence lifestyles. We have plenty of natural resources and we use these for food, medicine and building houses and canoes. We understand the importance of these local resources.

LAND RESOURCES

Land resources include plants, soil, domestic animals, wildlife, minerals, forest, swamps and wetlands.

1. Plants

Plant resources to meet basic human needs include the following:

Staple foods: banana, corn, sweet potato *(kaukau),* cassava, rice, sago, taro and yam

Legumes: peanuts, beans, peas, winged beans and soya beans

Edible greens or kumu: aibika, pumpkin tops, ferns, kangkong and cabbage

Fruits and nuts: avocado, breadfruit, citrus fruit, coconut, custard apple, galip nut, guava, mango, passionfruit, pawpaw, pineapple, soursop, pau and laulau

Beverages and stimulants: betel nut palm, betel nut pepper, cocoa, coffee, tea, sugar cane and tobacco

Spices: chillies, ginger, garlic, vanilla, cardamon and spring onions.

sugarcane sago pandamus

Non-food plants: vines, leaves, grasses, bamboo, rubber kapok, flowers and pyrethrum

For you to try

- As a class, walk around your school using the above list to identify the plant resources List these resources.
- Identify plant resources in your local area. List them according to abundance and usefulness.
- Identify plant resources that are for sale in your local area. List them and show how the sale of these resources benefits the local community.

Key facts

Climate and soil influence the type of vegetation grown in a certain area.

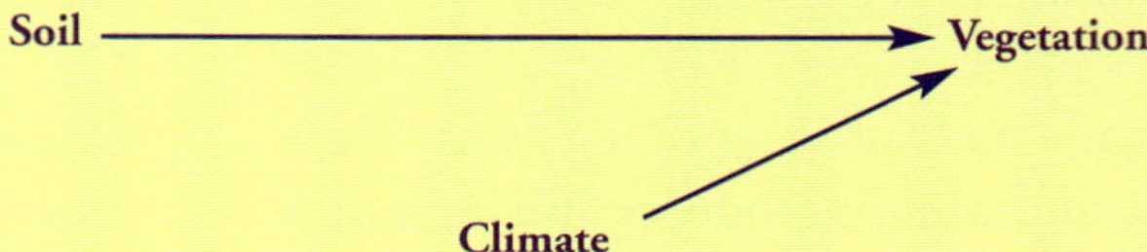

Looking at this diagram above. Where does soil come from?

Soil is formed when the rocks beneath the ground break up because of the interaction of water, heat and chemicals. So we can say that soils are influenced by climate (which provides the water and heat) and by geology (which provides the rock). Animals also influence soils: ants, worms and other creatures which live in the soil burrow through it, turn it over, expose it to the air and create chemicals which change the soil.

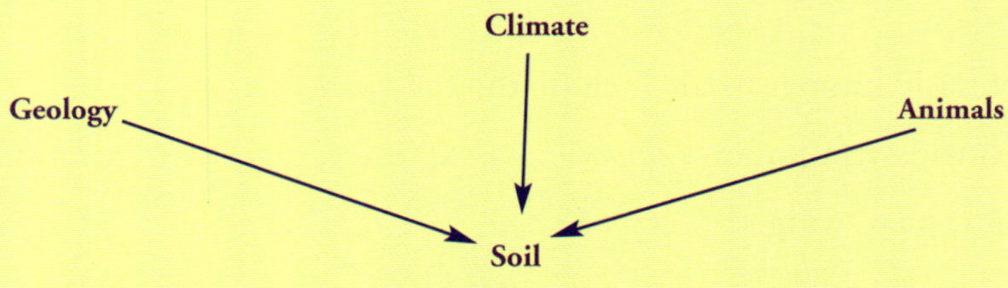

SS 2. Soil and land

The amount of land and the quality of the soil are important factors in how well land resources can be used to meet human needs. Where a family has a limited area of land, the same soil is used again and again. This will make the soil less fertile and less food will be harvested. The family needs to know ways of improving the quality of the soil.

1 All plants take food and water from the soil. The important foods are minerals such as:
   - nitrogen
   - phosphate (phosphorus)
   - potash (potassium)
   - lime (calcium)
   - magnesia (magnesium)
   - sulphur

2 Different plants need different amounts of minerals and the soil must suit the plant for it to grow well.

3 Fertile soils will have plenty of nutrients.

4 Adding organic matter such as mulch, compost and manure to soil helps to maintain fertility.

5 Rotating the types of crops grown in the same area also helps the soil remain fertile as different crops require different nutrients.

For you to try

- Walk around your school grounds and identify suitable land for food gardens. Mark out a garden bed. If no suitable land is available, find old wheelbarrows, drums, water tanks that can be filled with soil to make a garden.
- Visit community gardens in your local area. Ask people what they do to have fertile soil.
- Find containers that can be used as pot plants and fill with suitable soil.

3. Animals

Many domestic animals are kept by families for personal use, that is, food or for sale.

- **Poultry:** chickens, ducks, poultry products (eggs)
- **Livestock:** pigs, cattle, horses, goats, sheep, deer, water buffalo
- **Insects:** bees, butterfly

Wildlife can be a source of food, bilas, clothing or income for the family.

- **Birds:** pigeons, cassowaries, parrots
- **Mammals:** cuscus, tree kangaroos, bats, flying foxes, bandicoots
- **Reptiles:** snakes, crocodiles, turtles, eels
- **Insects:** grasshoppers, beetles, dragonflies, moths

For you to try

- Identify a range of animal resources (domestic and wild) in your local area.
- Go on an excursion to observe how domestic animals are housed and cared for.
- Write a report on how they are managed for both personal use and for sale.

4. Minerals

Exploration oil rig at Hides

In some areas of Papua New Guinea, there are mineral resources such as:

- gold
- copper
- oil
- nickel
- gas

These resources are non-renewable. This means that once we have used these resources they are gone forever as nature does not replace them. Mining these resources provides employment as well as revenue for land owners and the nation. Mining operations need to be managed properly as they can have a damaging effect on the environment and subsistence lifestyles.

For you to try

- Identify a mining activity in your area or in another province.
- List the positive and negative aspects of mining.

5. Forests, swamps and wetlands

Forests, swamps and wetlands are other resources. Many forests are rainforest. Rainforests are important because:

- Roots of trees protect the forest soil from rain.
- Falling leaves and fruit turn into plant food to keep the soil fertile.
- Fertile soil helps trees grow big and strong.
- Forest products are used by people for personal needs and to generate an income.
- Trees provide firewood and timber for housing, boats and carved items.
- Bark, grasses, fibres, vines and leaves are used for clothes, roofing and wrapping food for cooking.

Swamps and wetlands have land that is waterlogged and flat. They are important because:

- They are good homes for birds, fish and crocodiles.
- They provide excellent land for sago palms which are an important source of both food and building materials for people in that area.
- Rice growing is also possible in many wetland areas.

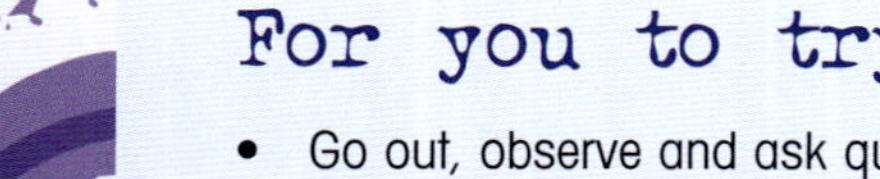

For you to try

- Go out, observe and ask questions for information on the use of resources from forests, swamps or wetlands. List these resources.
- In small groups, discuss how these resources are managed for personal use or for sale. Report your findings to the whole grade.

WATER RESOURCES 

In Papua New Guinea there often seems to be plenty of water, both as rainfall and in rivers. However, some areas have heavier rainfall than others and some parts of a year are drier than others.

Many people live near rivers, lakes or the sea because water is very important in their lives. Reasons include:

- Clean fresh water is essential for life and good health. It is used for drinking, cooking, washing and watering plants.
- Water can be stored in tanks, wells and dams. Supplies of fresh water need to be managed or there can be problems in dry seasons or times of drought.
- Rivers and seas are important for trade and travel.

Oceans, mangroves, coral reefs, rivers, creeks, swamps and lakes are all water resources. Marine life such as shellfish, eels, prawns, crabs, lobsters, sharks, squid, clamshells, coral, crocodiles and dugongs are water resources for personal use or for sale.

1. Marine life

Marine life refers to plants and animals that live in the sea, rivers and lakes. Plants and animals that live in this environment include fish, clamshells, corals, prawns, eels and seaweeds.

They live together and link with one another in the underwater environment.

Marine animals and their uses

Dugongs	used for meat (Now a national animal and protected since 1976. It is against the law to sell dugong meat. The animal can only be hunted using traditional weapons)
Fish	can be eaten and sold at the market for cash
Crayfish	used for food and export to overseas markets
Prawns	used for food and export to overseas markets
Clamshells	used for food and trade with other coastal villages
Eels	used for food
Corals	provide the home for many living creatures in the sea
Seaweeds	used for flavouring food, help in cleaning the digestive system, provide a natural paint for the canoes and brick houses in coastal areas.

For you to try

- Undertake a small research project to identify marine life in your local area (sea, river, mangroves or lake).
- Invite a guest speaker to talk about how marine life is used by your local community.

2. Oceans

The ocean covers nearly three-quarters of the surface of the Earth. It is home to many living creatures. In the ocean there is a natural life cycle; animals eat other animals to get their food. Oceans lose water through evaporation but most of the water lost returns as rain.

The ocean around Papua New Guinea is important because it allows big ships to carry people and cargo between Papua New Guinea and overseas countries. In addition, Papua New Guineans use the ocean to sail their canoes to trade yams, sago and clay pots to neighbouring islands and atolls.

Papua New Guineans also catch crocodiles for their skins and meat. They also collect shells from the ocean and use them to gain an income by making things like anklets, necklaces and headbands.

For you to try

- List resources gained from the ocean.
- Students living in coastal areas could collect resources such as shells or driftwood that could be used for craft activities.
- SS Use your atlas to name some of the big rivers that flow into the ocean around Papua New Guinea.

SS 3. Mangroves

Mangroves are coastal trees with long roots that hang over the sea and penetrate deep into the sea floor near the coastline, coral reefs and mouths of rivers. Mangroves provide shelter and are the best breeding place for fish, prawns and crabs. The trees hold water in their roots during low tide to keep the marine life under the roots alive. Mangroves are very hardy and can survive during drought and low tide periods. People use mangrove trees as a building material, especially for houses, because the material is very strong and can withstand the effects of sea water over a long period of time. The mangrove fruit is also useful as it can be splintered and boiled to extract salt for flavoring food.

For you to try

- Students living in coastal areas visit mangroves and observe and collect materials that can be used to make things.
- Invite someone with knowledge of mangroves to talk about mangrove resources and their uses.

4. Coral reefs

Coral

Coral reefs are made up of living organisms. Many of these coral reefs protect the coast line. Coral reefs are a good feeding ground for marine life. At low tide, people go out to the reef to harvest marine life such as fish, shells and coral crabs. Some coral reefs have small stones with white sandy beds. Many coastal children spend leisure time paddling small canoes, swimming, playing and diving on these reefs.

Some living creatures found on the coral reefs are:

- fish
- crabs
- eels
- crayfish
- clamshells
- star fish

For you to try

- Coastal students may visit reefs at low tide to observe and maybe catch marine life and later cook it.
- Look for information from libraries or newspapers on coral reefs in Papua New Guinea and how to protect them.

5. Fresh water

Fresh water is essential for life and there would be no living thing without it. Water falls from the sky as rain. It then forms rivers, lakes and seas and it is drunk by people and animals. If there was no more rain, plants and animals, including people, would die. Earth without water would be a very different planet.

Papua New Guinea is lucky because we have water all around us. There is water in the sea, on the land and in the air. The National Water Board is working hard to preserve and treat fresh water from rivers in large tanks to supply water to people living in towns and cities. People use water for drinking, laundry, gardening and cooking. In local community schools and villages, people store their fresh water in tanks and dug-out wells.

For you to try

- Identify the source of clean water supply for families and your school.
- Visit your local area and observe how people store their water at home. Draw a diagram to show how drinking water reaches the user.
- Survey people living in a settlement or working in a large plantation to find out how they obtain their water and who pays for it.
- Investigate how clean water is supplied to your school. Draw a diagram.
- Invite a guest speaker from the National Water Board or local community to explain how they manage water resources for the area.

6. Rivers, creeks, swamps, lakes

Rivers are water flows from the top of the mountains to the sea. Papua New Guinea has many large rivers. The rivers are important for the generation of electricity, transportation, fishing and farming. Rivers provide a good home for fish, prawns, crocodiles and birds.

Creeks are smaller than rivers. Creeks flow between valleys, rocks and mountains and then join main rivers. People use creeks for fishing, fetching water and laundry. Creeks are a good home for fish, prawns and eels.

Swamps are waterlogged soil. Swamps are large areas of land soaked in water. This land is not suitable for gardening. People use swamps to hunt crocodiles and wild deer. In Manus Province, Worei people use swamp land for planting and harvesting sago.

Lakes are pools of water with no outlets. Lakes are the preferred home and breeding place for crocodiles. People use lakes for canoe transport, fishing and laundry. Some lakes are very large and have small islands on them.

For you to try

- Go on an excursion to a nearby river, creek, swamp or lake to observe how people use this resource. For example: fishing, washing clothes, toileting. List uses and evaluate their environmental impact.
- Use your atlas to identify the major rivers and lakes in Papua New Guinea and describe how they help people make a living.
- Identify which provinces have large areas of swamp land and describe how the community uses the swamps to make a living.

HUMAN NEEDS PD

1. Physical needs

Every person has needs that must be met if they are to survive. These needs include:

- **Food:** People and living creatures need food to live, that is, our body needs nutritious food to survive. Without food we would die.
- **Shelter:** People need shelter to keep away the cold, rain and heat from the sun. Good shelter means healthy people and healthy living makes life easier.
- **Water:** All living things need water to survive. Water is essential for life. In some foreign countries, like Pakistan, people sell water on the street for cash.
- **Air:** Human beings, animals and plants need air to respire. Without air, life as we know it on this planet would not exist.
- **Clothing:** People need clothing for protection. Clothing can vary depending on the climate and society in which you live.

These five elements are essential to support human life on earth.

2. Other needs

Every person has other specific needs that enable them to function within society.

- **Love:** There is need for love within a family and society. In Papua New Guinea, the extended family shows great love and care from one to another. This love keeps people together as one big family.
- **Protection:** Everyone needs to feel safe and secure in their environment. In a caring and safe environment people live longer and are able to contribute to the welfare of family, parents and the community.
- **Safety:** People need to be safe in a work place, at home and at sea. In large mining areas, safety comes first, then work. Human lives are precious and safety at work is very important. People must wear protective clothing and footwear and comply with safety precautions at all times.

For you to try

- In small groups discuss your family's basic needs and how they are met. List these needs.
- Describe the differences in resources used by you and an ancestor to meet physical needs.

IMPORTANCE OF RESOURCES

The use of land and water resources must be managed wisely if we are to provide for the needs of people now and in the future. If these resources are mismanaged, human life is in danger.

1. Resources to support human life

The resources that directly support human life are:

- Food that can be gathered from forests, lakes, rivers and the sea. *Forest* – meat from animals, fruit from trees, nuts, materials for houses. *Lakes, rivers, sea* – marine life for food, making handcrafts and building materials.
- Plants that can be grown and harvested in gardens, e.g. kau kau, corn, abiba, bananas.
- Livestock that can be reared, e.g. chickens, pigs.

For you to try

- Make a list of food that can be collected from the forest, rivers, swamps, lakes and the sea.
- Visit a livestock project. Ask, observe and record what you see.

2. Resources as a source of livelihood

The following resources enable average Papua New Guineans to make a living:

- selling surplus food from gardens
- selling extra pigs and chickens from farms
- selling animal products, like eggs, from farms
- selling cash crops
- farming fresh-water fish and selling them to others in town
- selling honey from farms
- growing and selling vegetables
- making and selling handcrafts
- making and selling art work
- growing and selling plants
- organising tours for visitors to their community

For you to try

- Take your extra produce and sell it at the market.
- At the market, question sellers about successful selling strategies.

3. Valuing resources SS

Valued resources in Papua New Guinea include:

- handcrafts
- betel nut (traditionally important in PNG culture)
- kwila trees
- Bird of Paradise
- orchid plant
- vanilla plant
- balsa tree for timber (East New Britian)
- clay pots
- dogs' teeth
- traditional shell money

Bird of Paradise

For you to try

- Visit an elder in your community to talk about valued resources for ceremonies.
- A Make up a list of valued resources that can be found in your local area.
- Write an information report about items used for traditional dress. Draw these items.

APPROPRIATE WAYS OF MANAGING RESOURCES

1. Conservation

Conservation of resources means using them wisely. Conservation has always been important in the traditional life of Papua New Guinea. This has included:

- keeping soil fertile
- controlled hunting and fishing to allow wildlife to breed and multiply
- protected forest areas and water resources to allow people to live off the land

These practices have protected the natural resources of Papua New Guinea for thousands of years.

By conserving these resources, people can:

- produce goods
- satisfy their needs
- improve their standard of living

Resources must be managed so that they are protected for the future.

Some good conservation practices include:

- using the same garden land without cutting down more forest
- growing trees for firewood on grassland instead of cutting down forests
- growing trees for houses and furniture
- not lighting grass fires
- growing trees and plants that will attract wildlife such as birds

For you to try

- Consider appropriate ways to conserve the resources in your local area. For example, mulch gardens to conserve moisture in soils; build edges around a garden to protect plant life.

2. Reforestation

The word **reforestation** means the replanting of young trees after old trees have been harvested. In large forestry plantations, mature trees are harvested and young seedlings are replanted. This process is essential if Papua New Guinea is to maintain a sustainable timber industry. In addition to supporting a forest industry, reforestation helps to prevent soil erosion and allows the local animal and plant life to thrive.

Reforestation can be found in the Transgogol Valley in Madang Province. When the company harvests the mature timber, the area is cleared and new young timber trees are planted. These young trees take seven years to mature before they are harvested. This reforestation cycle is one of the Forestry Acts which the company must follow to satisfy the Forestry Law of Papua New Guinea.

For you to try

- Obtain tree seedlings or small trees and plant in a suitable site around your school or community.
- Role play different attitudes to logging from the point of view of a forest animal, company, land owner, employee, conservationist.

3. Sustainable fishing, hunting and gathering SS

Good sustainable practices involve taking only what you need for useful purposes.

Papua New Guineans need to develop sound procedures to sustain fishing, hunting and food gathering practices. Practices such as dynamite fishing, which kills more fish than necessary and damages the marine environment, and taking fruit from trees before it's ripe to use as missiles are wasteful and irresponsible. These procedures should ensure there will be resources for future use. In this way, people's physical needs are satisfied while safeguarding resources for the next generation.

For you to try

- Interview a local person about fishing and hunting practices. Consider how sustainable these practices are.

4. Coral reef protection

Coral reefs need to be protected from dynamite fishing and other practices. Harmful practices including excessive fishing, interference by humans and damage from boats often destroy marine life and affect reef environment. Starfish numbers need to be controlled, as starfish destroy young coral by sucking out the nutrients.

Community leaders must educate people so that they use the reef wisely to meet their needs as well as sustain it for future generations. Reef protection programs have already started in some coastal areas. For example: The West New Britain local group has established a viable coral reef program.

For you to try

- Students on the coast can visit coral reefs and assess their condition, and can identify sustainable practices. Be careful not to damage the reef.

SS 5. Protection of the natural environment

The natural environment must be protected and conservation laws enforced if it is to be preserved for future generations.

The natural environment will be protected if:

- Land owners are fully consulted about the use of their land, sea and underground exploration.

- National parliament members work closely with the people in deciding the best possible ways to conserve the environment and control damage.
- People form 'interest' groups and set rules to control the use of resources in their local environment.
- Provincial government takes measures to prevent the mismanagement of local resources and brings those mismanaging these resources to justice.

For you to try

- Investigate the natural environment in your local area. Discuss any mismanagement of natural resources and suggest better practices.

6. Crop rotation

Crop rotation is the planting of different crop types on the same piece of land each planting season. Crop rotation helps to keep the soil fertile and also helps minimise the number of diseases and pests in a field. Like vegetable crops, fields crops should be grown in rotation.

Monocropping means growing one crop type on the same land season after season. Monocropping is a bad farming practice. It is bad for the soil and for the crop. It makes the soil less fertile, so more fertilisers are needed to keep up the crop yield. Pests and diseases increase, which means more pesticides and labour have to be used. All these increase the cost of growing a crop and decrease the farmer's profit.

		Year			
		1	2	3	4
	A	Corn	Peanut	Cocoa	Grass
Field	B	Peanut	Cocoa	Grass	Corn
	C	Cocoa	Grass	Corn	Peanut
	D	Grass	Corn	Peanut	Cocoa

Example of crop rotation

For you to try

- Investigate crop rotation practices.

- Visit local gardens and create a timeline to show crops grown in the same garden at different times of the year.

7. Grazing management

Managing grazing practices of cattle requires planting different pastures on the same field.

The advantages of mixed pastures are:

- The pastures become established quicker.
- Better use is made of the land.
- Legumes help the grasses to grow by adding nitrogen to the soil.
- Production and nutritional value are increased.

For you to try

- If cattle are kept in your area, interview owners on grazing practices.

SS 8. Water conservation

Improved water practices such as water tanks, dams, mulching, avoiding dripping taps and keeping animal water clean are essential if we are to conserve water and make sure it is 'safe' for drinking.

The Water Board Treatment Plant's major objective is to treat and supply fresh water to people in specific, needy areas. We must assist the Board by respecting these properties and follow best management practices to conserve water.

Tolai fish trap, Rabaul Harbour

For you to try

- Implement simple ways of conserving water such as mulching gardens, using animal dishes above ground for drinking.

9. Wildlife management SS

Papua New Guinea needs to protect its abundant wildlife for future generations. Papua New Guinea has many beautiful birds, butterflies and animals and if not protected, they will be in danger of extinction. This requires sound management so that wildlife can flourish in their natural environment.

New methods of hunting should be discouraged as wildlife is killed unnecessarily. People should continue to use traditional hunting methods, which allow wildlife to thrive in their natural environment. Legislation is required to ensure the protection of wildlife, especially endangered species, and people need to be made well aware of the legislation and the penalties they will incur if they disregard it.

For you to try

A • Walk in the bush. Observe all wildlife, record what you see (for example, frogs, spiders, birds, butterflies). Group and record your findings in a table.

The Environment

Key facts

- The environment can be described as all the things which surround us.
- Some parts of the environment are NATURAL – mountains, river valley forests, the climate . . . Other parts are INFLUENCED BY PEOPLE – grassland, vegetation around settlements, soil erosion. And some parts are MADE BY PEOPLE – houses, airstrips and roads for example.
- People influence the environment and the environment influences peple. We call this INTERDEPENDENCE; people and the environment influence each other.
- The environment affects all people wherever they live.
- Things that happen in one place may significantly affect things in other places, even far away.

SS ELEMENTS OF A LOCAL ENVIRONMENT

Example: Elements of the environment at Tapini (Central Province)

Natural	Influenced by people	Made by people
Mountains	Grassland	Airstrip
River valley	Vegetation around Tapini	Houses
Forest on hill tops	Soil erosion on slope in middle	Tracks
Climate		Gardens

Tapini Village

For you to try

- Draw up a similar table to show the key elements of your local environment under the headings 'Natural' and 'Built'.
- A Make a 3D model of your environment making use of scrap materials such as sawdust, sand and clay. Indicate all elements of the environment.

1. Natural elements

The natural environment consists of:

- soils
- vegetation
- animals
- climate
- landforms

These elements influence each other. For example, the climate and soils of an area influence the type of vegetation which grows there. The type of landform and vegetation in an area influences the way human elements develop.

1 The different sorts of landforms may be found mixed together. You may find hills and plains together and plateaus and mountains together.

2 Papua New Guinea can be divided into regions where different sorts of landforms are common. For example:
   - Southern plains and lowlands – mainly plains with hills
   - Central ranges – mainly mountains with some basins and plateaus
   - Sepik-Ramu – almost all plains
   - Northern ranges – a mixture of all types
   - Islands – a mixture of all types

Key facts

There are four main types of landforms: Mountains, Hills, Plateaus and Plains.

Landform	Height	Shape	Slope	Relief
Mountains	High above sea level	Mainly sloping land	Tops may be gentle but sides are nearly always steep	Always large, i.e. there is always a big difference in height between the highest and lowest parts
Hills	Between 200m and about 1000m	More gently rounded than mountains	Usually not as steep as mountains	Medium
Plateaus	Can be at any height above 500m	Depends where they are: PIEDMONT – between a mountain and a plain INTERMONTANE – surrounded by mountains SUMMIT – on top of mountains	Always very gentle slope	Also very gentle relief
Plains	Height usually below 500m	Gentle with small hills	Very small slope	Very small relief

For you to try

- Using your atlas, look at how landforms are depicted on a map. Summarise this information and present your finding to your class.

- Write an information report on the landforms in your local area.

Vegetation

There are six main types of vegetation:

- **Mangrove swamp** – grows only in salty water usually between the levels of the tide. Mangroves prefer to grow in shallow, sheltered water. Very difficult for people to pass through.
- **Swamp** – found in wet areas not affected by the sea; areas wet because of climate or nearby rivers on flat land. Sometimes grows trees and palms, but mostly very tall grass.
- **Savannah** – a mixture of scattered trees, grasses and ferns.
- **Lowland forest** – very tall trees with much undergrowth; many different sorts of trees and plants and creepers. Usually in areas of high rainfall; a lot of animals, birds and insects. Rich in timber.
- **Mountain forest** – above 1000m the temperature is cooler and the forest in covered in mist and cloud most of the time. Moss is common, insects (especially ants and mosquitoes) and birds are less common. Trees shorter and thinner than the lowlands.
- **Mountain grassland** – above 2500m trees are usually small and few in number. Frost is common. The level that trees stop growing is called the Tree-line.

For you to try

- Walk around your area and observe all types of vegetation.
- Invite an elder to explain types of vegetation traditionally used by people in your area.
- Ask questions to discover: How much of the local vegetation has been cleared in your area? What is there in its place?

2. Built elements

People choose to live where they can satisfy their needs for shelter, food and water. This depends on the resources of the natural environment they live in. The climate and vegetation of an area often determine the type of houses people build, how they use the land and the type of crops they grow. The main resource required by all communities is water. People need water to drink, for cooking, washing and watering gardens.

For you to try

- Interview an elder of your local community and ask questions to find out why your village/town was built where it is?
- Consider the following:
  - shelter
  - transport
  - communications
  - energy (power)
  - leisure
  - education
  - water
  - clothing
  - food
  - culture/religion

Which are needs and which are wants?

- Walk around your local community and observe how many of these wants are evident.

3. Social elements

Communities differ in the ways they are organised. A community is not just a place in the environment. It is a place where people meet and interact with each other. People have **needs** (things they must have to survive) and **wants** (things they would like to make life easier). People meet their **needs** and satisfy their **wants** through activities with other people. Each person who belongs to a community has an important part to play in it.

Members of a community must:

- give and receive love
- protect and care for each other
- work together
- help and co-operate with each other
- respect each other
- learn to solve community problems peacefully

To be a healthy and happy community, the community should have:

- safe houses
- clean water
- enough healthy food
- work
- health Services
- education

A community will have many problems if people:

- are lazy
- are given too many free things
- fight each other
- do not have work

Elements of the local environment at Tapini

Natural	Built	Social
Plants	Bridges	Friendships
Animals	Roads	Community groups
Air	Farms	Family
Water	Village	School
Soil	Hospital	Clans
Sunlight	Church	Youth Club
Mountain	Airstrip	Women's fellowship group

For you to try

- Walk around your local area and observe natural and human built elements.
- Draw up a table for your local area, identifying natural, human built and social environmental elements.

CONSEQUENCES OF CHANGE IN AN ENVIRONMENT

Heavy flooding washed away bridges, schools, roads and three villages at the weekend

The two bridges that linked Madang and Lae Highway were washed away by heavy flooding at the weekend. The Sausi bridge over the Boko river, the longest Bailey bridge on the Ramu-Madang Highway, was destroyed, cutting off all access by road into and out of Madang. Two power pylons also collapsed, disrupting supply from the Ramu Yonki power grid to Madang. A large section of the Monia bridge, a few kilometers from Sausi, fell into the Homia river on Monday, making it impassable.

The estimated cost of the damaged is yet to be established.

The local people were heavily affected and government services like schools, health were put on hold due to heavy rain in the area. Although workmen have been sent to mend the bridges, this could take some months before things are back to normal. Many million tonnes of sediment and rich soil were washed away and deposited at the mouth of these rivers.

For you to try

- Discuss the above extract from the *National* newspaper dated 24 March 2004. The report shows how a natural resource can affect physical and human resources.
- If your school is in a rural area, make a list of the ways people or nature influence your environment.
- If your school is in an urban area, make a list of ways the environment (climate, soils, hills, vegetation . . .) influences the way people live in your town. Remember to include things that may **influence** other people a long way away.

The following explain what could take place when humans destroy a natural environment.

1. Loss of habitat 

A habitat is a specific area where plants grow and animals live. Loss of habitat leads to extinction of certain species of animals. For example, the wild tree kangaroo in Papua New Guinea is almost extinct. The cassowary population and white cuscus are decreasing in number. Traditional law should be enforced to stop the clearing of large areas for logging and development. Logging operations have destroyed some of the best habitat sites for these animals.

The Government should monitor the removal of special hybrid plants (like eagle wood, balsa and kwila trees and traditional herbs) which are sold overseas.

For you to try

- Invite an elder to your class and ask him/her to talk about wildlife habitats in the past.
- Go out into the bush and observe how small creatures such as frogs, spiders, birds live in their habitate – e.g. their food supply, how they make their shelter.

2. Extinction 

As more and more vegetation is cleared there is a real threat that many species of animals, insects and birdlife will become extinct. The big blue butterfly is almost extinct due to large oil palm and logging operations in the Sangara area of the Oro Province that have destroyed much of their habitat. It appears that money is more important than the preservation of Papua New Guinea's major symbols such as the world's biggest butterfly and the beautiful Bird of Paradise.

3. Introduced species into local ecosystems 

Some 'introduced' species of animals and insects invade local ecosystems and destroy plants and other animal species. For example, coffee rust, a fungal disease, damages coffee leaves and can kill young coffee trees. It has created its own ecosystem and must now be controlled within the environment.

4. Introduced diseases – plants and animals

Any introduced plants and animals entering Papua New Guinea must be checked by the quarantine officials. Farmers must be on the look out and report any outbreak of diseases such as foot and mouth disease in livestock. This must be reported to agricultural officers so that the disease does not spread to other parts of the country.

5. Deforestation

Deforestation comes from harvesting forest timbers without replanting. This is bad management practice and leads to soil erosion and the pollution of creeks and rivers. Deforestation is a serious problem in West New Britain and in Madang Province. Legislation is required to ensure no further deforestation occurs in these areas.

For you to try

- Write a letter to the Foresty Department to ask for free small tree plants.
- Plant the trees around your school grounds.

6. Soil erosion

Soil erosion means the loss of soil from grasslands, river-banks, hillsides and cultivated fields. Soil erosion is caused by

- wind blowing soil away
- water washing soil away
- sun burning soil when trees and plants are cleared away
- fires that destroy soil nutrients

Logging development also causes soil erosion. When logs are harvested, revenue is earned but soil erosion occurs. It is essential to immediately replant areas that have been logged and so keep soil erosion to a minimum.

A shortage of good land for farming also contributes to soil erosion. Vegetation in the drier parts of the country takes a long time to grow back. People do not always give vegetation enough time to grow back before clearing the land again.

For you to try

- Walk around your school gardens and school grounds to observe soil erosion.
- Use mulch and other materials to prevent further erosion.

7. Land disputes

The destruction of traditional hunting grounds causes land disputes. As the population grows, more land is required for gardening and so land disputes are becoming common. People wish to earn more money, so they need more land for cash crops. Local farmers keep many pigs so they need more land to grow food to feed the animals.

In recent times, oil palm and mining industries have created major land disputes amongst tribal groups.

8. Introduction of foreign animals and plants SS

Papua New Guinea must be cautious when introducing foreign animals and plants into the country. Many countries throughout the world are still experiencing problems with 'introduced' species, for example, the rabbit problem in Australia. The Papua New Guinean Government must take precautions before allowing the importation of new animals and plants into the country.

For you to try

- Invite an officer from the Agriculture Department to give a talk about quarantine laws in Papua New Guinea.
- Go on an excursion into the bush around your local community and look for plants and animals that may have been introduced into Papua New Guinea.

9. Modification of the environment SS

Modification of the environment relates to new developments that actually change the environment. Some developments can produce revenue while not harming the environment, for example, reforestation. However, many other modifications lead to major environmental changes that cannot be reversed, for example, mining activities.

ENVIRONMENTALLY FRIENDLY ACTIONS

There are many good ways of practising environmentally friendly actions in the everyday life of your school and local community.

1. Waste management practices

Recycling of used papers

- Use the other side of printed paper for new work.
- Two-sided used printed paper can be used for art work.
- Collect and recycle all used papers in the school by selling the paper to garages for spray painting and panel beating purposes.

Composting

How to make a compost heap

- Clear an area 2.5m x 1.5m.
- Put down some cut grass or leaves.
- Put some peelings on top of this.
- Cover over with a bit of manure.
- Cover over the manure with 5cm of soil.
- Sprinkle ashes on top.
- Start over again with grass about 30cm deep. Continue as above.
- Add layers until about 1.2m high. Last, cover with soil or grass.
- It must be kept moist.
- It can be covered if there is too much rain.
- In three or four weeks, turn the heap so outside materials are on the inside.
- In two or three months it will be ready.

Compost can be made from any of these materials

Collect any kitchen peelings, rotten vegetables, chicken manure or leftovers from the market and allow these to decay. Organic materials become food to help plants grow and produce a high yield.

Reusing waste materials

Reusing old tyres to make flower pots

- Select medium-sized used tyres.
- Cut tyre in half.
- Turn the inside part out.
- Fold it in to a flower vase shape.
- Use a nail to insert holes through the tyre and fasten the edges with a strip of copper wire.
- Close the bottom part with a piece of tyre or sheet metal.
- Paint the outside and add pictures.
- Add soil and ornamental plants and put them for sale.

Reusing old tyres to protect the flower garden

- Collect 20 medium-size used tyres.
- Paint them white.
- Dig small shallow drains around your house near the flower garden. Place the tyres in the shallow drains vertically. It adds extra beauty to your residence.

2. Crop protection (Biological control)

There are some plants that can be used to discourage insects from destroying your vegetables.

- Marigold flowers, planted between vegetable plots, keep insects away. The flower produces a specific scent that disturbs grasshoppers from feeding.
- Chillies can also be used as part of an integrated pest management system.

Natural pest spray

- Collect 10 ripe chillies.
- Cover your hands with plastic bags and cut the chillies into pieces.
- Place these chillies into a bucket of 5 litres of water.
- Add 1 litre of water to another dish and wash your hands with soap. Then add this soap solution into the bucket containing the chillies. This solution is ready to be used on the vegetables.
- Use an empty fish tin with holes inserted at the bottom (use a nail). This will be your special cup for pouring the solution from the bucket on to the vegetables. Remember one cup for two or three plants depending on the size.

Note: The soap solution keeps the chilli's flavour on the leaves and grasshoppers will not do any more damage.

3. Soil Conservation practices

Growing legumes

Growing legumes helps to enrich the soil and improve soil structure. This group of plants have root nodules where the beneficial bacteria called **Rhizobium** live. The Rhizobium bacteria help to make plant food (nitrogen) in the soil.

Some legume plants are:

- peanut
- beans
- wing beans
- leucaena
- soya bean
- crotalaria
- puereria

Legumes put nitrogen back into the soil. Peanuts should be grown wherever possible, as they increase the nitrogen content of the soil. It's a good idea to plant vegetables like cabbage and aibika in the place where peanuts are grown as these vegetables benefit from the nitrogen and produce a high yield.

Crop rotation

Crop rotation means planting different crops on different plots each season. Why should we practise crop rotation?

- It helps to keep the soil fertile (full of nutrients).
- It helps to control pests and diseases.
- It is a good way of growing better crops without having to spend money on fertiliser and pest control products.

To understand crop rotation, you need to know that there are four main types of vegetables.

- **Root** vegetables (carrot, potato and wing-beanroot)
- **Leaf** vegetables (aibika, cabbage, kangkong, lettuce and water cress)
- **Fruit** vegetables (tomato, egg-plant, corn, okra, cucumber and pumpkin)
- **Legume** (peanuts, beans and peas)

Each group takes different nutrients from the soil. For example:

root vegetables take phosphorus out of the soil when they grow;

leaf vegetables take nitrogen out of the soil when they grow.

Every crop puts different nutrients back in to the soil. When you dig in the old leaves, stalks and roots, these feed the soil.

SS Contouring

A contour is a line joining all points of equal height above sea level. Your Geography Atlas has maps with contour lines to show the shape and height of hills and mountains. These lines are not usually straight. They go around the hill-side.

You know that whenever a sloping field is ploughed, there is a danger of soil erosion. Most farmers know that this danger is reduced if they plough across the slope, and not up and down. The plough furrows (small drains) will hold back the rain water, if they are exactly level, that is, if they follow the contour of the land.

Laying out contours

There are no marks on the ground to tell us where the contours lie, but it is not difficult to find them. A simple way is to use a piece of string joined to two upright poles about 5 metres apart. This apparatus is known as an 'A frame'. The string must be fixed at exactly the same height at a marked point on each pole. The height chosen should be approximately eye-level, to make the apparatus easy to use. A special line level is hung on the string to show when it is exactly horizontal.

Honey farming

Honey is the fluid produced by bees from the nectar of blossoms from flowering plants. The bees collect the nectar and combine it with substances of their own. They then store the fluid and leave it to mature in honey combs.

Bees are kept mainly for the honey they produce. Beekeeping provides box hives for the storage of honey and this makes collection of honey easier.

Types of bees

Papua New Guinea has its own local bees. These bees make combs out of clay under roofs, tables – anywhere there is shelter.

The 'introduced' specie of bee is superior to local bees because its colony is normally strong and readily produces honey.

4. Energy saving measures

Key facts

Water conservation practices include:

- Store water in tanks.
- Repair leaking taps.
- Install water meters to control water usage.
- Dig and cement underground wells.
- Store fresh water in containers for village use when the stream is a distance away.

Energy saving practices include:

- Use lighting when required.
- Discourage any illegal connection of power for private use at home.
- Avoid overuse of power energy in private residences.
- Look after all installation equipment.
- Report any problem to the power authority concerned.

Note: People must understand that nothing is free. We must pay for the power and other facilities provided to us. These amenities are there to improve our living and lifestyle.

Solar energy 

Papua New Guinea is already using solar energy. Solar energy means receiving energy from the sun into battery cells where it is transformed into power to give us light, cook food, do laundry and to operate the television and radio sets. It is clean and does not create pollution. This is a cost efficient way of obtaining energy. Solar panels come in different sizes according to need. The bigger the need, the more solar panels required. The battery and other attachments are included in the purchase price.

5. Marine protection

People must take initiative to look after their coral reefs and lakes. The Mahonia Nadari of West New Britain Province is a community that looks after their marine areas. The community sets guidelines. Anyone operating outside these guidelines, for example, by illegally fishing or diving within these set zones, is prosecuted.

The Mahonia na dari run programs for other institutions on basic diving and marine management. The centre is very popular and has attracted many marine scientists from around the globe. This is a good example of a community protecting and preserving their resources while earning a living. The centre is being used as a model and training institution for good management of coral reefs and marine life.

Crop and Animal Management

ECONOMIC VALUES

Economics is the study of the management of money and resources. Some important economic principles apply directly to agriculture.

First, let us look at the study of **supply, demand** and **price** at work.

These terms describe the forces at work in a market in which goods of any kind are bought and sold. The 'supply of goods' means how much there is for sale. The demand for a good means how keen people are to buy it. In a market the price paid is determined by the balance between supply and demand. If the supply is short and the demand is strong, the price will be high.

Farmers know that if they can supply something scarce to a market (supply is low) they can get a high price, because everyone wants what they are selling. Early crops often command a high price for this reason. Later in the season, when supply is plentiful, the demand weakens and the price drops.

The table below shows how the action of supply, demand and price operate at a market situation.

Madang market and crop season

Type of crop	Number of farmers involved	Supply	Demand	Price
Watermelon	Few in July Many in November and December	Less High	High Less	High Less
Corn	Few in November Many in December and February	Less High	High Less	High Less
Cucumber	Few in November Many in December and February	Less High	High Less	High Less
Taro	Few in July Many in January and April	Less High	High Less	High Less
Oranges	Few in June Many in July and August	Less High	High Less	High Less
Banana	Few in August and October Many in December and March	Less High	High Less	High Less
Pawpaw	Few in July and August Many in February and August	Less High	High Less	High Less

Crops have economic-value in many ways. They:

- provide revenue for the family, province and the nation
- improve the condition of the soil
- provide nutritious food for both humans and livestock
- help in research work
- help in the production of medicines
- assist in manufacturing fertilisers and other chemicals to control insects

Animals carry the following economic value. They:

- provide income for an individual, family and province
- help in the export sector and bring revenue for the country
- provide nutritious food (protein) for the family
- help in research work
- assist in the production of tinned meat (Oxen palm and Globe meat)

For you to try

- Visit your local markets and list the types of crops sold.
- Write a report about what you observed about supply, demand and cost of crops.

PD NUTRITIONAL VALUES

Table A: *Carbohydrates and Fats (energy foods)*

Food Group	Food	Percentage of carbohydrate	Percentage of fat
Cereals	Rice, polished	46	1.7
Beans	Sago	88	0
	Corn	74	4
	Snake bean	62	1
Others	Soya bean	10	1.7
	Sugar, white	99	0
	Onion, red	74	0
	Honey	69	0
	Milk, sweetened, tinned	55	8

Note: Percentage of carbohydrates and fats in 100 grams of certain uncooked food

Key facts 

- Carbohydrates are solid substances that are made up from carbon, hydrogen and oxygen.
- Foods rich in carbohydrates are the main source of energy for living things.
- A simple diet of rice provides a large proportion of the carbohydrates required.
- Rice and sago contain large amounts of starch which is used by the body to produce energy and heat.
- Sugar is also fuel for the body's energy and heat production. It can be eaten in small amounts as too much sugar in the diet leads to weight gain as well as tooth decay.
- Fat is found in small quantities in nearly all foods. In the body it is broken down to produce heat and energy.
- It is important to eat some fatty foods because of the vitamins they contain.

Table B: *Protein (basic health foods)*

Food group	Food	Percentage of protein
Animal protein	Chicken	20
	Fish, fresh	18
	Egg, fresh	13
	Beef, tinned	24
	Fish, tinned	21
	Pork	15
Vegetable protein	Soybean	36
	Peanut	23
	Long bean	23
	Onion, dried	12
	Rice, slightly milled	8

Note: Percentage of protein as in 100 grams of certain uncooked food

Key facts 

- Proteins are body-building substances found in all living organisms.
- No food is totally protein. There is also a water and fat content.
- Most people need about 60 grams of protein per day. Children require about 80 grams per day as they are still growing and need a larger amount of these body-building foods.
- Milled rice supplies about 10 grams per meal or roughly 30 grams per day. If people eat rice each day, they will obtain half of their daily protein requirements.
- Foods that have little protein include tapioca, kaukau, banana and yams.

Table C: *Composition of foods commonly used in Papua New Guinea*

Nutrients in 100 g of edible portion	Water (g)	Energy (joules)	Protein (g)	Fat (g)	Carbo-hydrate (g)	Fibre (g)	Calcium (mg)	Iron (mg)	Vit. A (I.U.)	Thiamin (mg)	Ribo-flavin (mg)	Niacin (mg)	Vit. C (mg)
BODY-BUILDING FOODS													
Beans, dry	10	1418	24.0	1.7	57	4.0	110	8.0	–	0.50	0.20	2.00	–
Eggs	74	661	13.0	11.5	1	–	55	2.8	1000	0.12	0.35	0.10	–
Fish, fresh, fillet	81	305	17.0	0.5	–	–	20	0.7	–	0.05	0.10	2.50	–
Fish, tinned	51	1314	22.0	24.0	1	–	44	13.0	100	0.06	0.20	2.60	–
Liver	70	569	20.0	4.0	5	–	10	10.0	20000	0.30	2.50	13.00	30
Meat, fresh, beef	61	1096	16.0	22.0	–	–	10	2.5	–	0.07	0.15	4.50	–
Meat, tinned	59	950	23.0	15.0	–	–	15	4.0	–	–	0.20	3.50	–
Milk, full-cream powder	4	2092	25.5	27.5	38	–	900	0.8	1200	0.30	1.15	0.76	13
Pandanus nuts, dry	9	2800	11.9	66.0	–	–	–	–	–	–	–	–	–
Peanuts, dry	6	2422	27.0	45.0	17	1.0	50	2.5	–	0.90	0.15	17.00	–
Winged beans, dry	12	1636	30.6	18.3	26	9.4	215	18.0	–	1.03	0.33	3.09	–

Key facts 

- Vitamins are organic compounds that people need in their diet for healthy growth.
- International units (i.u.) are a measurement used for measuring vitamins.
- Each unit is an extremely small quantity.
- A person needs about 25 mg of vitamin C every day.
- Vitamin D is known as the sunshine vitamin as the human body has the ability to produce vitamin D with the aid of sunshine.
- Foods which are rich in vitamin A also contain vitamin D.

Table D: *Protective foods table*

Nutrients in 100 g of edible portion PROTECTIVE FOODS (Fruits)	Water (g)	Energy (joules)	Protein (g)	Fat (g)	Carbo-hydrate (g)	Fibre (g)	Calcium (mg)	Iron (mg)	Vit. A (I.U.)	Thiamin (mg)	Ribo-flavin (mg)	Niacin (mg)	Vit. C (mg)
Avocado pear	75	690	1.5	15.0	6	1.5	10	1.0	200	0.07	0.15	1.00	15
Banana, ripe, sweet	70	485	1.0	0.3	27	0.3	7	0.5	100	0.05	0.05	0.70	10
Guava	80	243	1.0	0.4	13	5.5	15	1.0	200	0.05	0.04	1.00	200
Orange	86	222	0.8	–	13	0.3	30	0.5	30	0.08	0.03	0.20	45
Mango	83	264	0.5	–	15	0.8	10	0.5	600	0.03	0.04	0.30	30
Passionfruit	75	385	2.3	2.0	16	3.5	10	1.0	20	–	0.10	1.50	20
Pawpaw	89	163	0.6	–	9	0.7	20	0.5	1000	0.03	0.03	0.20	50
Pineapple	85	238	0.4	–	14	0.8	20	0.5	100	0.08	0.03	0.10	30

Key facts

- The body requires certain nutrients (minerals) to be healthy and stay alive.
- A person needs about 1 gram of calcium every day.
- Salt is found in every food, especially dried or salted fish or other sea foods. It is also available in table form from stores.
- Iodine can be found in most water which comes from wells or springs because the water releases iodine from the rocks.
- Iodine is found in all sea foods and iodised salt.

For you to try

- Students bring one food (any kind) from home. Group the foods into food groups. Discuss the nutritious value of all the food.
- Go to the market and observe local foods that are sold. List all foods and identify the food group to which they belong. Record this information as a chart.
- Using this chart, plan a weekly menu to ensure family meals are balanced.

CULTURAL VALUES

In the past, crops and animals have promoted traditional values and identity. Many gardening systems in Papua New Guinea have traditionally been able to meet the nutritional needs of the community. People have grown staple foods in their gardens, collected vegetables, fruit and nuts in the forest and hunted birds and animals. All these activities provided a good and balanced diet for the people.

Today, in many parts of Papua New Guinea, people are being forced to change ways of obtaining food. This is because the bush – an important source of food – is further from the village or town.

Gardening in Papua New Guinea is changing and so are the types of food people now eat.

For you to try

- Invite an elder to your school and ask him or her to describe food gathering during their parents' times. Has it changed? How? Has it improved the quality of life? Were there any foods that had a cultural value and why?
- Go on an excursion into the bush and try to find foods identified by the elder. Use these foods as people did long ago.

Province	Type of food or animal	Cultural value
Highlands	Kaukau Pigs Cassowary	Bride price payment Traditional singsing Occasions, moka activities and clan compensation
Coastal & Island	Sago Yam Fresh fish Pigs	Bride price payment Traditional singsing Occasions, and clan compensation

Strings of fish at Koki

IMPORTANCE OF CROPS AND ANIMALS

1. Keeping livestock

There are many important reasons for keeping livestock on a school farm or garden.

Keeping livestock enables primary school students to gain practical experience in livestock management and care, such as feeding, housing, breeding, health care, processing and storage, sale of products, finances and record keeping.

Animals, and their products, can provide food and income for the school. Animal waste can also be used to improve the school garden. Raising livestock on school grounds can also encourage closer school and community relationships. However, there are some issues you need to consider before deciding to raise livestock at the school. Thorough planning and preparation need to be undertaken before animals are brought to the school.

For you to try

- Go on an excursion to a nearby farm with livestock, e.g. ducks, chickens. Ask the farmer questions on how to set up a small school livestock business. Write a report outlining how this small business can be started.
- Complete the animal diagram below.

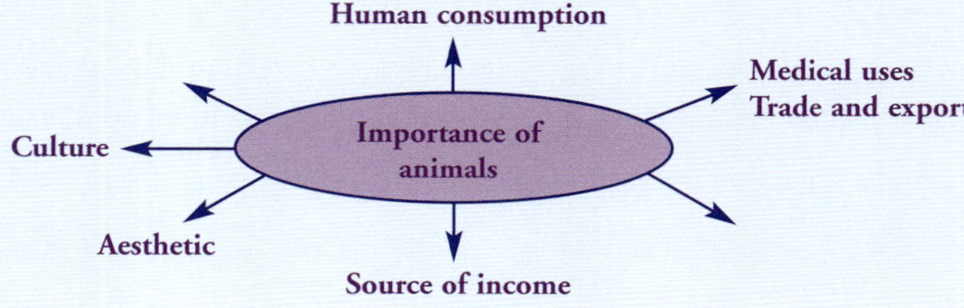

2. Growing crops

The vast majority of Papua New Guineans practise agricultural gardening. Many of these gardeners use traditional techniques that are suited to the local environment. It is important that new techniques for growing crops are used alongside traditional practices to produce more food for a growing population.

It is important to increase our understanding and knowledge of growing crops and this can be done in a school setting. Students must learn gardening skills. Skills such as digging rows and mounds, draining, planting, mulching, manuring, harrowing, weeding, thinning, fertilising and staking.

Growing pumpkin

For you to try

- Invite a village gardener to talk to the class about how to make a school or backyard garden.
- Work in your school garden to practise the gardening skills being taught in class. Keep a record of your activities in your journal.

3. Human consumption

Rural people still grow their own food and are mostly self-sufficient. They grow food, keep animals, gather and hunt. This kind of subsistence farming has been used for many generations. In the past they used a barter system of trade to exchange goods for things they could not produce themselves. Now, money is more widely used throughout Papua New Guinea, and villagers receive money in exchange for their produce. They use this money to purchase goods they need and cannot make.

For you to try

- Visit a traditional farm in a village and observe how the farm supplies all the needs of the people. Write a report.
- Ask questions to find out how they plan their work so that they can make sure they have everything they need.

4. Source of income

In recent years, new ways of farming are being introduced, as people learn to apply science to agriculture. You are learning about new ways in school. People sell their surplus crops and animal meats at market to earn an income. With an income, they can buy rice and tinned fish from the store to supplement their diet. Some earn money to pay children's school fees or to buy new clothes.

For you to try

- Go into the local town or large village and observe how people make an income such as selling handcrafts, food, catching fish, selling ice-blocks.
-  Ask people questions on how much money they can earn in a week. Write a report on your findings.

5. Trade and export

In many countries subsistence farmers are learning about modern agriculture and are trying to implement modern practices. Growing cocoa, coffee, rubber, vanilla, balsa trees, vegetables and fish farming represents one kind of modern agriculture. Feeding pigs and poultry in house is another kind. Modern farmers grow their crops or animals to sell for money. To make money they need to grow many plants and rear many animals on their land so that they have enough to feed their families as well as selling produce at market. Many of their products are traded within the country and some are exported to overseas markets to increase the country's revenue.

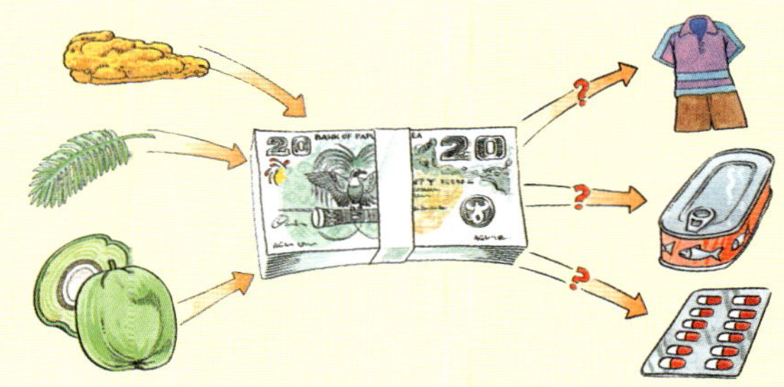

Resources sold for money to buy goods

For you to try

- Invite a successful business person, plantation owner or fishing company representative to talk to your class about what is required to set up large businesses for export. Ask questions and write up a report.

6. Medicinal uses

Subsistence and modern farmers grow crops and animals for their own needs. They also use special crops for medical and ceremonial use. Special animals are chosen and kept aside for traditional family worship and other social obligations. Traditional practices are good because they benefit the people within the community.

For you to try

- Invite an elder to explain how they use plants or animals to heal sick people.
- Invite a local man or woman to talk to the class about herbs that students can use to heal sores and swollen limbs.
- Go into the bush near your school and collect some of the plants that can be used to heal sores. Give these plants to the local people.

7. Aesthetics and cultural uses

In many societies in Papua New Guinea, traditional herbs are used to celebrate cultural ceremonies. Some herbs are used for healing and others for matrimonial purposes. The New Ireland and East New Britain Provinces are famous for their use of herbs. Read the experience of one person.

> I was at Duke of York Island and I wanted to get to Kokopo on the mainland. There was a big storm and the local people asked me to wait. They made a fire, burnt 'special' leaves and the villagers danced around it. Within twenty minutes, the storm had disappeared and the people rubbed the remaining leaves on me to make my limbs and neck flexible and tough for the long journey. On my journey to the mainland the sea was calm and the sky blue. It was truly a remarkable experience.

TRADITIONAL CROPS AND ANIMALS

Subsistence agriculture was the basis of the traditional economy. Many traditional crops and animals are significant in our societies.

Benefits of traditional crops include:

- resistance to disease
- suited to local conditions
- seed collection and propagation materials are always available
- maintain cultural heritage and identity

- Promote and maintain traditional farming methods
- Maintain local biodiversity

For you to try

- Walk around your local community and make a list of traditional crops which are valuable and can be used for weddings and other traditional ceremonies.
- Identify animals in your local area that are highly valued for traditional ceremonies, singsing and compensation payment.
- If your school is in the highlands, invite an elder to talk about moka celebrations. Make a small item that can be used in these celebrations.

1. Resistant to diseases

Important crops believed to have originated in Papua New Guinea are resistant to most diseases. These are the winged bean, sugar cane and two native tree crops – the Hoop pine and Klinki pine.

Current trends in Papua New Guinea are to increase food production through different cropping and farming systems, by using high yielding disease-resistant crop varieties.

For you to try

- Invite an agriculture officer to your school to talk about crops that are resistant to diseases in your local community. Write a report and draw up a chart showing these crops.

2. Seed collection and propagation

Farmers need to choose seed carefully. They need to collect good quality seed varieties that are recommended for the area. Good quality seeds are not mixed with dirt. They are free from disease and weedseeds and have a high germination rate.

The word **propagation** describes a new plant that grows from a part of a single parent. In this case, there is just one parent plant and no seeds are sown. The new plant has all the characteristics of the original parent (shape, color, height and taste). They cannot be different as they are really the same plant. Variations between generations are not possible.

For you to try

- Visit gardens around your school. Identify plants that can be propogated, e.g. cassava, pumpkin, taro.

- Invite an agriculture officer to demonstrate the techniques of propagation. Practise these techniques in your school gardens.

3. Maintenance of cultural heritage and identity

People in Papua New Guinea are very proud. They have grown food in their gardens, collected vegetables, fruits and nuts in the forests and hunted birds and animals. All these activities provide a good diet. The skill in growing yams, kaukau, banana, sago, winged bean, pitpit and sugar cane is a cultural heritage and part of the identity of this country. The yam harvest with traditional singsing signifies happiness and joy within the community for a successful harvest. Gardening is the key to our culture, heritage and identity.

For you to try

- Learn songs and dances used in harvesting celebrations in your local area.
- Sing these songs during the School Open Day activities.

4. Maintenance of traditional farming methods

Many traditional farming methods are used for survival and pleasure. Any new agricultural systems must co-exist with these traditions. As Papua New Guinea develops, people will use the system that suits them best. Papua New Guinea is a big country and no one system suits all our needs. We should maintain some traditional farming methods used by our ancestors such as subsistence farming because they require less labor and are inexpensive to operate.

Yam tuber plant

5. Maintenance of local biodiversity

Biodiversity is the existence of a large number of different kinds of animals and plants that make a balanced environment. Ecosystems are communities of living things that interact to support each other. Sunshine, rain, soil, plants, animals and humans are all linked together in ways that enable living things to survive. Nature provides this biodiversity and people need to ensure that it is protected. We need to be educated to care for and maintain local ecosystems. We need to learn good management practices and become responsible owners of our environment.

For you to try

- Design posters as part of a school awareness campaign on proper management of our forests, coral reefs, rivers, lakes and swamps.

COMMERCIAL CROPS AND ANIMALS

Agriculture in Papua New Guinea is a way of life. For over 30,000 years our people have practised some form of farming and gardening. Today 85 per cent of people live in rural areas and depend on farming.

Few farmers own tree crops such as cocoa, oil palm, coffee, rubber or spice crops as they do not have the capital required to run such enterprises. However, some farmers grow vegetables, tuber crops and sugar cane for extra cash and others plant balsa and kamarere trees for timbers. Others look after farm animals for traditional reasons and for commercial export to overseas markets.

Farmers can now choose from different ways of farming. Also, farmers should combine different farming enterprises because if one crop fails, another may succeed.

The following farming enterprises can provide food for the family and also produce surplus which the farmers can sell. This is known as commercial agriculture.

Kama village coffee garden

Enterprise	Produce
Vegetable production	Cabbages, beans, onions, tomatoes
Crop production	Kaukau, potatoes, corn, cassava
Poultry production	Broilers, eggs, hens
Cattle rearing	Milk, beef, hides
Cash crops	Cocoa, coconut, oil palm, coffee, vanilla

1. High yield

Farmers aim to make a profit on produce grown. To do so, farmers need to ensure they produce a high yield. They need to apply good farming practices.

For you to try

- Ask the people in your community how they calculate the yields of their gardens and farms.
- Harvest your school garden plots and work out how much produce you have grown and what income has been generated.

2. Fast growth

Farmers select good healthy crops and animals that grow fast. This is the aim of the business. The early maturing of farm products brings success and pride to the business. High returns are cost effective.

Planning and management of the whole farm organisation, combined with quality output, is essential to farming success.

3. Commercial market

The **commercial market** means the market for cash crops like coffee, cocoa, copra, fish, timbers, spice crops, fruit and minerals. As these crops are produced in commercial quantities, profits gained are very high. The sale of these crops brings in revenue for companies and the government.

Other commercial market sales include:

- Shell – used for decoration and shell money.
- Crocodiles skins – these are bred in farms and are an important source of money for village people because their skins sell for high prices at commercial markets overseas.
- Fishing – there is some danger that the number of fish in the waters around Papua New Guinea will become smaller in the near future. Anticipating this, in 1979 the government claimed the fishing rights to the area of sea up to 325 kilometres from the coastline. Countries who wish to fish in these waters must give payment to the government.

Salting crocodile skin

For you to try

- Organise a visit to a large scale commercial enterprise and observe practices. Ask questions. Write a report.
- Use the world map to name countries that import our cocoa, coffee, oil palm, fish and timbers.

4. Export potential

Orchid plants, traditional bilums and oil palm seeds have great export potential. This is because they are unique and can only be found in Papua New Guinea. Local communities are looking for other products that can also be exported. Gold, oil, nickel, vanilla and rice are products that have great export potential.

For you to try

- Invite a skilled craftman to show you how to make craft items that may have export potential – necklaces, anklets, jewellery.
- Make one item and sell at the market.

COMPARE BENEFITS

It is important that new skills in farming, new ways of making gardens or looking after animals be compared with existing techniques. In so doing, we can improve farming.

The main reasons we compare the benefits of traditional and commercial farming include:

- to improve yields
- for fast growth and early sale
- for community needs to support traditional activities
- to supply enough planting materials for the new gardens
- to meet high demand for food from people at the market or in town

Growing sweet potatoes in the highlands

Some of the traditional benefits of crops and animals are:

- to support compensation payment
- to promote family status within the community
- to support the wealth of a community
- resistant to diseases
- to maintain closer relationship with the neighbouring clans
- to sustain barter systems of trading in rural village settings
- to provide and support human consumption

Some of the commercial benefits of crops and animals are:

- To provide a source of income for families and the country
- To use for trade and export
- To provide good quality seeds and that means increased yields
- To develop cross breeding programs for new breeds of animals suited to conditions in Papua New Guinea
- To develop new varieties of crops for rural farmers to grow for extra cash

For you to try

- Conduct a survey to identify valuable crops and animals in your local area.
- Hold a class discussion on the benefits of both traditional and commercial farming.
- Hold a class debate about which farming system would be most beneficial and appropriate to your local area.

Better Living

Chapter Summary

In this chapter you will have an opportunity to:

- ✔ **Investigate the effects of healthy and unhealthy lifestyle practices.**
- ✔ **Investigate effective management skills relating to home and school environments.**
- ✔ **Discuss the need for effective money management, record keeping and budget planning. Identify a situation (at school or home) where these skills can be applied in a practical way.**
- ✔ **Plan and design things to make; for personal use or to generate an income.**

Syllabus references

Strand: Better Living

Substrand 1: Healthy Living

Outcome: 6.2.1 Describe and demonstrate aspects of personal hygiene, nutrition and safety that promote a healthy lifestyle.

Substrand 2: Care and Management

Outcome: 6.2.2 Develop an awareness of the importance of effective management skills in creating pleasant and safe environments at home and school and apply and make suggestions for improvement.

Substrand 3: Wise Consumer

Outcome: 6.2.3 Describe the importance of effective money management, record keeping and budget planning and apply this to a practical situation at home or school.

Substrand 4: Making Things

Outcome: 6.2.4 Identify resources and a range of tools and equipment required for handcrafts and apply the skills in a familiar practical situation.

Healthy Living

HEALTHY LIFESTYLE

Papua New Guinea needs people who know how to have a healthy lifestyle. This includes factors such as living conditions, exercise, personal and community hygiene and nutrition.

1. Living conditions

What is the condition of the classroom in which you learn?
What is the condition of the community in which you live?
What is the condition of the house in which you live?
Is it healthy or unhealthy?
What makes it a healthy environment?
What makes it an unhealthy environment?

Healthy living conditions include the following:

- clean, not dirty, free of pests
- neat and tidy, not messy or cluttered
- well maintained, not in need of repair
- airy and well ventilated
- a happy and safe community where people care for one another
- enough space for individuals, not overcrowded

A For you to try

- Using the criteria above, describe the condition of your classroom.
- Identify any shortcomings.
- Design a plan of action to fix any problems.
- Implement your plan as far as you can.
- Evaluate the results.

2. Physical exercise PD

Exercise or physical activity such as swimming or walking is part of having a healthy lifestyle.
Do you have plenty of energy or do you often feel too tired to do things?
What physical activities do you do day by day?

A person with a healthy lifestyle leads an active life:

- going to school and back
- joining in class activities
- doing housework
- doing gardening
- playing with friends
- playing games
- looking after animals

For you to try

- Investigate the type of physical exercise you do.
- Plan and design a personal physical fitness program.
- Implement it for a month and then evaluate the program.

PD 3. Personal and community hygiene

Personal hygiene is essential for people to have a healthy lifestyle.

This includes:

- an easily accessible, plentiful supply of water for washing purposes
- people with knowledge and skills in keeping bodies, clothes and homes clean
- people with knowledge of germs and harmful substances and dirty habits to be avoided
- people with good habits in keeping themselves, their belongings and their environment clean
- clean habits in preparing and eating food
- knowledge of appropriate and hygienic toilet use and care
- knowledge of appropriate and hygienic ways to dispose of rubbish
- knowledge of appropriate and hygienic ways to keep animals

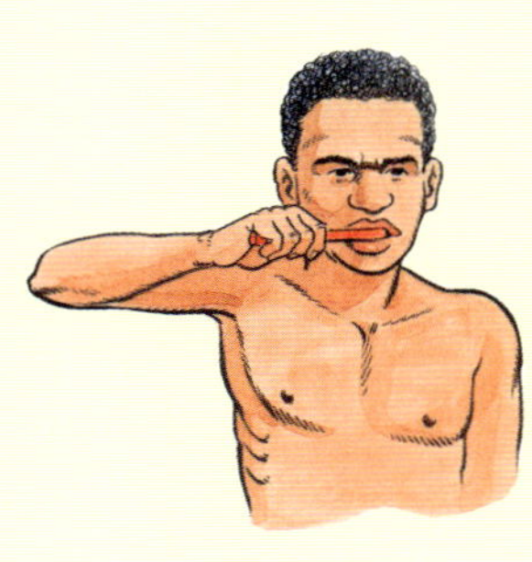

For you to try

- Describe and demonstrate how you care for your skin, hands, feet, hair, ears, eyes and nose.
- Identify a problem to do with personal hygiene and suggest ways to solve the problem.
- Design a poster to promote clean habits.

PD 4. Nutrition

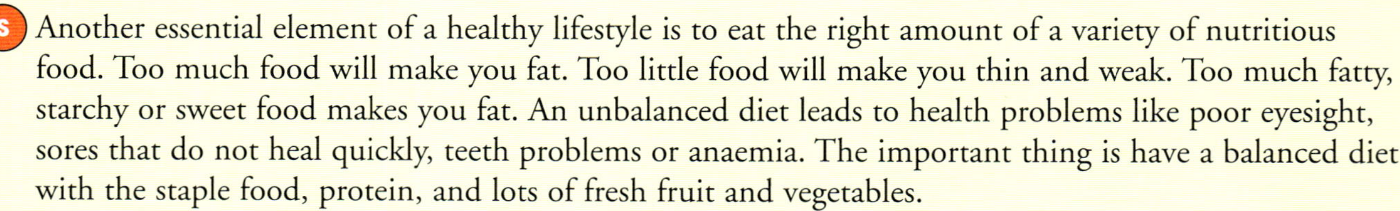

S Another essential element of a healthy lifestyle is to eat the right amount of a variety of nutritious food. Too much food will make you fat. Too little food will make you thin and weak. Too much fatty, starchy or sweet food makes you fat. An unbalanced diet leads to health problems like poor eyesight, sores that do not heal quickly, teeth problems or anaemia. The important thing is have a balanced diet with the staple food, protein, and lots of fresh fruit and vegetables.

For you to try

- Conduct interviews to gather evidence on the food intake of a classmate in the past 24 hours.
- If this were a typical day in that person's life, make a judgment on whether or not that person is getting enough nutritious food.

PERSONAL HYGIENE

As individuals, we are responsible for keeping ourselves and the places where we live clean. We need knowledge and skills for personal hygiene and we need to feel proud to be clean. Hygiene is important to kill germs and control the spread of diseases.

1. Washing the body

Remove all clothes. Wet the body all over. Rub soap all over the body. Pay particular attention to dirty spots and body creases. Rinse off the soapy water. Dry the body using a clean bath-towel. Clean and cover any cuts or sores to prevent infection.

Washing hair

Wet hair. Rub in a little shampoo. Rinse off the soapy water. Comb and let dry. If head-lice are a problem, use a head-lice killing solution as directed.

For you to try

- Observe a baby or young child being washed.
- Imagine you are teaching a younger brother or sister how to wash his or her body. List the steps in the process that you would teach them.
- Imagine that you will offer a hair-washing service in your community. Describe and demonstrate how you would wash someone's hair. What equipment would you need? How much would you charge for the service? What other hairdressing services could be offered?

2. Caring for clothes

Washing clothes

Collect clothes to be washed. Sort clothes in the order to be washed, for example, whites, lightly soiled coloured clothes, very dirty clothes . . . Soak clothes to loosen dirt. Use bar soap to rub on dirty marks. Wring out soapy water. Rinse clothes in clean water. Wring to remove excess water. Peg clothes on a clothes line. (Usually clothes are hung the way they are worn.) When clothes are dry, unpeg, fold and store in a clean, insect-free place.

Washing machines can also be used to wash clothes.

Ironing clothes

It is important to know about types of fabric if clothes are to be ironed. A hot iron can melt and spoil clothes made from nylon or some other synthetic fabrics. Use a cool to warm iron on synthetic fabrics. Use a warm to hot iron for rayon and cotton fabrics. If unsure, test by ironing a facing, seam or hem on the wrong side. Creases are removed more easily if a steam iron is used or clothes are sprinkled with water before ironing. When ironing clothes, it is usual to iron the small parts of a garment first. This includes collars, sleeves, pockets and hem edges before ironing the larger parts of a garment.

Mending clothes

To keep clothes attractive and to make them last a long time, mending should be done as soon as it is needed. This could mean re-stitching seams or hems, darning tears or patching holes.

Removing stains

Bleach is a useful product for removing stains from white clothes. Use according to directions on the bottle. Lemon juice is a useful substitute for bleaching some stains out of clothes.

For you to try

Investigate whether or not you could earn an income by offering a Clothes Care Service to children or teachers at your school or to people in the wider community.

- Plan or design a project. It could be a clothes washing, ironing or mending service.
- Consider the resources you would need and fees to be charged.
- Design an advertisement.
- Implement your plan for a term and then evaluate its success.

3. Caring for the house

Read the following and then plan an income generating house-cleaning service.

Floors

- Remove any rubbish.
- Sweep.
- Mop.
- Vacuum.

Walls and ceilings

- Dust from top downwards to remove loose dust.
- Wash surfaces with detergent in water.
- Use a cleaning powder or paste on mould or hard dirt.
- Rinse with clean water and wipe dry with a cloth.

Windows

- Dust to remove loose dirt.
- Wash with detergent in water.
- Use a cleaning powder or paste on hard dirt.
- Rinse with clean water and wipe dry.
- Newspaper helps to dry windows and leaves glass clear and smear free.

Kitchen

- Collect, sort and stack equipment to be washed.
- Prepare washing up water with detergent.
- Wash equipment from the least dirty to the most dirty, for example, from cutlery and glassware to cooking pots.
- Use a scouring pad with cleaning powder or paste to remove hard dirt or grease.
- Rinse equipment.
- Dry and store in a clean, pest-free place.
- Wipe clean the table or bench tops and other kitchen surfaces.
- Sweep and mop floor.
- Dispose appropriately of any kitchen rubbish.
- Burn burnable rubbish.
- Give food scraps to animals.
- Put vegetable peelings on compost heap.
- Before throwing something out, consider reusing or recycling it.
- Place tins and bottles in rubbish bin or bury in rubbish hole.
- Recycle plastic bags.

Bathroom, shower, handbasin, toilet

- Use disinfectant in water.
- Wipe surfaces of shower, handbasin and toilet with a damp cloth.
- Apply a cleaning powder or paste on mould or hard dirt.
- Use a toilet brush to clean inside toilet bowl.
- Rinse surfaces and wipe clean.

Bedroom, bed, mirror

- Make bed by straightening, smoothing and tucking in sheets.
- Plump the pillow and arrange neatly.
- For clean mirror, wash surface with a damp soapy cloth; rinse off soapy water; dry with newspaper to leave a clear, smear-free finish.

For you to try

Imagine that you want to earn an income from cleaning people's houses.

- What would be your hourly rate?
- What equipment would you need?
- What knowledge and skills do you need?
- Choose a name for your house-cleaning business.
- Design an advertisement.
- List and cost the resources you would need to start your business.
- Design a cleaning schedule.
- If possible, ask someone to let you clean their house to try out your skills. Calculate the fee you would charge according to the time and number of cleaners involved. Ask the customer to evaluate the service.
- List the skills for caring for different types of houses in villages, settlements and towns. Explore house cleaning equipment and materials available in stores.

4. Cleaning the environment

It is difficult for people to have healthy bodies and healthy lifestyles if they live or work in overcrowded, messy, dirty buildings in need of repair that are surrounded by broken vehicles and with rubbish littering the ground. A clean environment promotes a healthy body and healthy lifestyle. It is up to you as an individual to do the right thing in caring for your environment.

An environment encourages us to have a healthy body and healthy lifestyle. We must make a personal promise to ourselves to have clean habits and to keep our surroundings clean and healthy.

A clean environment promotes a healthy lifestyle

We need to:

- Dispose of rubbish thoughtfully.
- Recycle or reuse items where possible.
- Put rubbish in rubbish bins, if they are available. If rubbish bins are not available, use other methods.
- Bury bottles, tins and plastic bags and containers in rubbish pits.
- Burn papers and leaves.
- Put organic matter like animal droppings, grass and plant cuttings in compost heaps.
- Give food scraps to animals.
- Deal with problems (bad smells, ants, flies and rats) in appropriate ways.

Rubbish bin

If drainage is a problem, talk about ways to solve the problem.

- Where is the waste water coming from?
- Look at the soil type and slope of the land. Consider the need for stone drains, pipe drains or ditches.
- Think about what we can do to solve a drainage problem.

For you to try

- Identify ways to have a clean environment in the classroom, school grounds and surrounding community.
  Compose slogans and make posters to encourage people to keep the environment clean.
  Make and decorate a rubbish box for the classroom.
- Identify ways to re-use items that we would otherwise throw away.
  Make flower vases out of the bottles.
  Make pencil holders out of tins.
  Make sawdust stoves out of tins.
  Use old cartons for storage.
  Design and make useful things by reusing plastic, flour and rice bags.
- Draw a picture of an unhealthy environment.
  Label all the things in the picture that make it an unhealthy environment.
  Show your picture to another person.
  Explain why it represents an unhealthy environment and what people should do to turn it into a healthy environment.

ASPECTS OF GOOD NUTRITION

PD
S

1. Food groups

Key facts

Knowing about nutrition is knowing about the substances in food and how the body uses them for energy, growth and protection from illness.

Important nutrients in foods are CARBOHYDRATES, FATS, PROTEIN, VITAMINS and MINERALS. These can be called the five food groups.

Different foods contain different amounts of nutrients. Food like milk and eggs contain all the nutrients. Food like cassava and sago mainly contain the carbohydrate – starch.

We need to know about nutrients so that we can plan to eat nutritious meals and snacks.

The best way to know about the nutrients in food is to study food composition tables. Fish and meat are good sources of protein. But they do not only contain protein. They contain other nutrients like fat, iron, calcium and vitamin B.

Eating a mixture of different foods is healthy. A mixed diet provides the body with different nutrients.

Fats and carbohydrates are used by the body for energy and warmth.

- Fats provide twice as much energy as carbohydrates, but we usually only eat fats in small amounts.
- There are three types of carbohydrates – starch, sugar and fibre.
- Sugar is found in sweet foods like sugar cane and fruits.
- Fresh fruit and vegetables contain fibre and this is important to aid digestion and disposal of waste from the body.
- The foods we eat in the largest quantity are starchy food like rice, sweet potato, taro, yam, bananas, sago, cassava, and flour products like biscuits and bread.
- Everyone has a 'staple' food and this is the main starchy food that is eaten by an individual, family or community.

Protein is used by the body for growth and repair of body tissues 

- Protein foods are especially important for children from birth to adulthood. These are their growing years.
- Protein is rich in animal foods and vegetables such as peas, beans and nuts.
- Animal tissue is similar to human tissue and because of this animal protein can be more readily used by the body than vegetable protein.
- Animal protein foods include all meats (chicken, pork, lamb, beef), seafood (fish, crabs, eels) and animal products (milk, eggs, cheese).

There are many vitamins and they are known by letters of the alphabet

- **Vitamin A** is good for healthy eyes and skin. It is found in carrots, pawpaw, guavas, avocados, mangoes and dark-green leaves.
- **Vitamin B** works to release energy from energy foods and is found in a mixed diet.
- **Vitamin C** helps build up the body's resistance to infection and is good for healing tissue. It is found in foods like lemon, orange, tomato, pineapple and leafy vegetables.
- **Vitamin D** is good for healthy bones and teeth. Sources of vitamin D include sunlight, oily fish, eggs, liver and milk.

Important minerals are iron, iodine, calcium and phosphorus

- **Iron** is needed for healthy blood and preventing anaemia. Iron is found in fish, meat, nuts, wholegrain cereals and dark-green leafy vegetables.
- **Iodine** is needed for a healthy thyroid gland and prevents goitre. It is found in iodised salt and seafood.
- **Calcium and phosphorus** are good for healthy bones and teeth. They are found in a mixed diet, green leaves and milk.

A For you to try

- Role-play being a particular nutrient.
- Dress-up as a type of food containing that nutrient.
- Introduce yourself as that nutrient and explain to an audience why they should eat you and what would happen to them if they did not eat you.
- Practise your role and then put on a play for another class. Be entertaining, but be sure to get your message across.

PD 2. Balanced meals

We need to eat meals that are nutritionally balanced to have a healthy body.

- Balanced meals are meals that have some food from different food groups. An example is fish (for growth) with rice (for energy) and aibika and pumpkin (for protection).
- Nutritious snacks are also important. Putting egg or meat and tomato on a buttered biscuit is more nutritious than simply having a biscuit.

Bread, lettuce, tomato, meat, butter

Try these recipes. They are all balanced meals.

Stick Food

You need

Thin sticks 20 cm long
Sweet potato
Carrot
Onion or capsicum
Pineapple
Fish or meat

You do

1. Prepare thin sticks, for example, from ribs of leaves on coconut fronds.
2. Cut food into very small pieces.
3. Push food onto each stick in a repeating pattern.
4. Cook by grilling over hot coals or on an oiled barbecue plate.

Extra: More flavour can be added by sprinkling food with garlic, chilli, ginger, soy sauce, curry or coconut cream.

Fresh Meat Stew

You need

Meat, chicken or fish
Taro or sweet potato
Onions
Pumpkin, carrot or tomato
Green leaves such as aibika or pumpkin tops
Salt
Water or coconut cream

You do

1. Cut food into small cubes or bite-sized pieces.
2. Put all food except the green leaves into a large saucepan and add liquid until food is almost covered. Add salt to taste. Cook until food is tender.
3. Add the green leaves and cook a further five minutes. Serve.

Extra: Garlic, chilli, ginger or curry can be added for more flavour.

Fruity Ice Cups

You need

1 small ripe pawpaw
2 ripe bananas
juice from 1 lemon
juice from 1 green coconut
paper cups

You do

1. Mash pawpaw and bananas.
2. Mix well with juice from a lemon and a green coconut.
3. Pour into paper cups.
4. Place in a freezer until frozen.

For you to try

- Plan an income-generating project selling nutritious snacks to school children during morning tea or lunch breaks.
  Investigate what children like to eat and can afford to buy.
  Plan and design what to make.
  Produce your items.
  Market them and evaluate the project.
- Start a class recipe book with recipes tried in class as well as recipes collected from family and friends.

3. Methods of food preparation

- Moist heat methods of cooking include boiling, stewing, steaming and mumu-ing. When cooking in these ways water or coconut cream is added to the food to be cooked.
- Dry heat methods of cooking include frying, baking, grilling and smoking.
- Food is prepared in different ways to give variety and to prevent food presentation from becoming boring.

Try these recipes that need different cooking methods.

Boiled Eggs

You need
Eggs
Water

You do
1. Place the eggs in water in a saucepan.
2. For a hard-boiled egg, keep in boiling water for at least five minutes.

Extra: Hard boiled eggs can be mashed with a little milk, coconut cream or mayonnaise to become creamy. They are delicious served with lettuce on buttered bread, biscuits or on slices of cooked taro or sweet potato.

Baked Banana and Sago or Cassava Wrap

You need
Banana leaves
Ripe bananas
Sago or Cassava
Coconut cream

You do
1. Prepare leaf wraps by softening banana leaves over hot coals and cutting into squares.
2. Grate cassava or (if using sago) rub sago until fine.
3. Mix mashed banana and coconut cream into the cassava or sago.
4. Wrap mixture in banana leaf squares and tie into parcels with vine.
5. Bake over hot coals or in an oven for approximately one hour.

Extra: Delicious if served with additional coconut cream on top.

Sausage Sandwiches

You need
Sausages
Bread slices
Margarine
Lettuce
Sliced tomato
Tomato sauce

You do
1. Grill, barbecue, boil or fry sausages.
2. Spread margarine on slices of bread.
3. Place a lettuce leaf, tomato slice and sausage on a slice of bread. Add a little tomato sauce. Cover with a second slice of bread.

For you to try

- Experiment and try different methods of preparing food for healthy living.
- Plan income-generating projects and use different food preparation methods for nutritious snacks to sell to school children and teachers.
- Implement and evaluate projects.

PD SAFETY

Safety at school, at home and travelling to other places is important for a healthy lifestyle. Safety means to be alive and well, out of harm's way, protected, undamaged, unharmed, unhurt, not injured and free from danger.

1. Being safety conscious

To be safe it is important for you to be aware of when you are vulnerable or at risk. You need to be aware of when you and your belongings are in danger.

Unsafe situations include:

- walking by yourself
- leaving your house or car unlocked
- standing or sitting on the edge of a boat or on the back of a vehicle
- leaving a fire or burning candle unattended
- holding large sums of money which others can see

It is not possible to list all the things that can be done to keep yourself, your belongings, your home and your environment safe. The following are just a few examples.

1 To walk with friends is safer than walking alone.
2 Lock your house or car when it is unoccupied.
3 Sit in vehicles and boats wearing a seat belt or life jacket so that you don't get hurt if the driver stops suddenly or a big wave hits the boat.
4 Place lit candles in a safe manner so that they cannot cause a fire if you fall asleep or the wind blows them over.
5 Handle money discreetly so that others do not easily notice it.

For you to try

- Are you conscious of times you could be unsafe? List five situations in which you feel unsafe. For each situation, what can you do to lessen or avoid being at risk?
- Interview another person and find out when they feel unsafe. It could be riding in an unroadworthy vehicle, climbing trees, walking on a busy road or visiting a big town. There are lots of possible situations. What advice would you give those people to minimise the risk?
- Make a collection of newspaper cuttings which report law and order problems. Talk about actions that could have been taken to keep people safe in those situations?

2. Storage of chemicals (S)

There are many things in a home environment that are poisonous and can cause death if misused. They need to be stored carefully.

Tablets for health problems must be kept where young children cannot reach them. Children may think that the tablets are lollies. If you notice a child eating anti-malarial tablets, stop them immediately. Get the child to vomit them up and take the child to a health centre quickly.

Kerosene, bleach, disinfectants and petrol are examples of dangerous liquids around homes. They are poisonous if swallowed. Children may think they are soft drinks. Do not store kerosene, oil or petrol in soft-drink cans or bottles. Petrol and kerosene are also dangerous as they are flammable and can cause fire. Children should not be allowed to play with matches. These dangerous substances need to be stored in a locked or secure place. If someone swallows poisonous liquids, do not try to make them vomit. Instead, get them to drink lots of water or milk to dilute the dangerous liquid and get the person to a health centre quickly.

Rat and cockroach baits, ant-killing powders and fertilisers contain chemicals that are poisonous if taken into the body. Children may touch these and then eat food with unwashed hands. They will become sick because traces of chemicals enter their bodies. Fertilisers and pest-killing substances need to be stored safely out of children's reach.

A For you to try

- Visit places in the school or community where harmful substances are kept. Talk about measures taken to keep them out of the way of people who might misuse them.
- Design a safety poster for storage of harmful substances.
- Role-play a situation that gives a safety message.

3. Safe storage of equipment

Guns, knives, axes, scissors and tools need to be handled safely and stored in an appropriate place. They are not toys for play. When carrying or passing sharp equipment, keep the sharp edges away from the other person. They cause cuts and injuries if misused or not kept in safe places.

Care needs to be taken around steaming kettles and hot pots of food to avoid getting burnt. Young children need to be watched closely when they are near stoves or fireplaces.

Check cords, plugs and power points for damage before using electrical equipment. Do not use them if they are damaged. Take care with electrical equipment to avoid electrocution, electric shocks and death. Never poke things into electrical wall sockets. Do not use appliances with frayed electrical cords or damaged plugs. Do not go near electrical wires that fall down during storms and look out for wires amongst trees that you climb.

For you to try

- Observe and demonstrate correct handling of equipment to avoid accidents.
- Visit places in your school or community to see where equipment is stored. Undertake a project to store equipment neatly and safely.

PD 4. Safety measures

Accidents happen quickly and easily. Accidents can happen anywhere. The best way to avoid accidents is to behave in a thoughtful and responsible manner. Always be conscious of ways accidents occur.

- Climb trees carefully taking care not to fall.
- Look carefully where you are walking so that you do not stand on sharp objects or get bitten by a snake.
- Walk on the right hand side of the road.
- Walk carefully up and down steps and on wet surfaces to avoid slipping and falling.
- Pick up broken glass, tins, nails and sharp pieces of metal and place them in a rubbish hole or rubbish bin.
- Pay games with safety in mind so that others do not get hit by a bat or tackled in a dangerous manner.
- Mend loose edges of mats and keep floors clear to avoid tripping over.
- Swim in safe places. Avoid strong currents. Don't dive into shallow water. Look out for sharks, crocodiles or dangerous things in the water.

For you to try

- Conduct a survey of accidents that have happened to your classmates or members of their families.
- Write down the causes of the accidents.
- Suggest ways the accidents could have been prevented.
- Role-play situations to give a safety message.

Care and Management

SAFE HOME AND SCHOOL ENVIRONMENT PD

1. Clean and accident free

A pleasant and safe home and school environment needs to be clean and accident free.

Dust, dirt, mould, spider webs, empty tins, broken glass, old bottles, plastic rubbish, bits of paper, leaves, plant cuttings and untidy storage of rubbish and personal belongings all make our environment look unclean and untidy. Accidents can happen by tripping over rubbish or loose floor mats, slipping on wet ground surfaces, or by standing on nails or bits of broken glass or tins.

What can we do about it? Dust, sweep and clean surfaces regularly. Remove spider webs as soon as they are noticed. Pick up and promptly dispose of rubbish in a thoughtful manner. Store belongings in a neat and safe manner. Mend or remove things that might cause accidents. Display signs to remind others to keep areas clean and safe.

2. Comfortable, attractive, inviting and pleasant

Think about your classroom and school. What makes your school environment comfortable, attractive, inviting and pleasant?

- enough space for the number of people
- suitable furniture and facilities for activities to be done
- attractive trees, gardens and play areas
- paths or walkways to get from one place to another
- well-maintained buildings
- supply of clean drinking water
- signs or notice boards with relevant information
- good ventilation and air flow in buildings
- clean, safe and free from rubbish
- interesting activities and friendly people

For you to try

- Investigate ways to make your classroom and school environment more comfortable, attractive, inviting and pleasant.
- A Plan a project to improve some aspect of your classroom or school. Make flower beds. Decorate the interior of a room. Landscape around the school. Dispose of rubbish in appropriate places.
- Carry out the project and evaluate the result.

3. Free from pests

Which pests are found in our home and school environments and how can we minimise the problems they cause?

Common pests in our environment are mosquitoes, ants, flies, cockroaches and rats. Insect spray to repel or kill insects is effective but expensive. Do you know of effective alternative methods?

Mosquitoes annoy us because they disturb our sleep with their buzzing and bite; they also cause malaria. Mosquitoes breed in stagnant water, so get rid of places where mosquitoes can breed such as old tins and containers. Sleep under mosquito nets. Burn mosquito coils to repel mosquitoes.

Ants can be a problem both inside and outside the home. There are ant killing powders and sprays that can be bought. Cleanliness and careful storage of household items can reduce the problem of ants in the home. Wipe up spills of food quickly so they do not attract ants. Tins of water can be placed under legs of food storage cupboards to prevent ants crawling up the legs and onto food. Follow ant trails outside of the home to find ant nests. These can be destroyed by fire or boiling water.

Flies, cockroaches and rats breed on food, dirty rubbish and animal excreta. We can lessen the problems they cause by keeping food covered, storing food carefully, disposing of rubbish in suitable ways and burying animal excreta. Cockroach baits, rat poison and rat traps can be used to trap these pests. Care should be taken not to put them where babies or pets move about.

For you to try

- Identify a pest problem in your school or community.
- Talk about a plan of action to reduce the problem.
- Implement the pest control plan.
- Report and evaluate the results.
- Investigate use of local materials to repel insects.

4. Neat and tidy furniture arrangement

Furniture includes movable things like beds, cupboards, tables, chairs, desks and bookshelves. What guidelines would you suggest for a neat and tidy arrangement of furniture in our homes and school?

- Furniture should be in a suitable position to serve its intended purpose.
- Furniture should not block access for people to get from one area to another.
- Individual pieces should look good in the whole arrangement of a room.
- Lighting is an important factor in placing tables and desks in areas where work is to be done.
- Changing the arrangement of furniture from time to time adds interest and variety to living conditions.

For you to try

- Draw up plans to show three different ways to arrange the furniture in your classroom. Implement a different plan each week for three weeks. In the fourth week, evaluate which furniture arrangement worked best.
- Draw up a floor plan for your family home. Show the furniture arrangement in each room. Suggest ways to change the placement of furniture and give reasons for the suggested changes.

EFFECTIVE MANAGEMENT SKILLS

1. Save money, time and labour

How can we save money, time and labour in carrying out daily activities?

Save money by keeping things in good order. In this way they will last longer before needing to be replaced.

Save money by using available resources instead of shop-bought goods. Coconut husks and rough leaves can be used instead of steel wool. Bamboo or other bush material can be used for fencing instead of chain wire. Bush vines can be used instead of rope. Brooms can be made from spines of coconut fronds. Room dividers and blinds can be made from local materials such as cane, bamboo, shells, pandanus and vines. Tyres and household containers can also be recycled for useful purposes. Rakes can be made from bamboo. What other useful things can be made?

Save time and effort by doing cleaning, tidying and mending jobs in an ongoing way when tasks are small and easy. Don't wait until jobs become large and difficult; it then takes more time and effort to clean, tidy and mend.

For you to try

- Identify things you can make from available resources to help in your work, living and play areas.
- Design a product. It could be for personal use or to sell.
- Obtain the materials and make it.
- Use the item and take pride in its appearance and usefulness.

2. Reduce accidents PD

A neat, tidy and well-cared for environment reduces the chance of accidents happening. It is important for each person to notice things or behaviour that could cause an accident and to take appropriate action. Each person needs to be responsible for keeping his or her area accident free.

Wet surfaces cause accidents that can be avoided. Take care walking or running on wet surfaces. Wipe up wet spills promptly. Put up signs, 'Danger, wet surface'.

Sharp objects cause accidents that can be avoided. Pick up broken glass, tins, nails and old pieces of metal.

Loose steps, broken floorboards and other damaged items need to be repaired quickly before an accident happens.

Talk about other causes of accidents and ways they can be avoided.

For you to try

- Investigate your school or community environment. Identify a situation where an accident could occur.
- Create a plan of action to make the situation safer.
- Implement the plan and write a short report about what was done.

3. Work efficiently with limited resources

An efficient worker is someone who plans ahead to get work done quickly and properly without wasting time. This gives the person more time to relax and do other activities.

Efficient people:

- Use equipment safely
- Use appropriate skills and methods
- Are creative in using available resources to improvise equipment
- Have a plan for getting jobs done quickly and properly
- Get other people to help with specific tasks
- Have a place for everything and keep everything in its place, when not in use
- Are economical in using resources such as soap, cleaning agents and disinfectant
- Care for limited resources such as brooms and bush knives to prolong their usefulness

For you to try

- Identify a job that you feel is not being done efficiently or properly at your school or in the local community. It could be grass cutting, rubbish collection or care of an ablution block, a rubbish disposal area, classroom or playing area.
- Produce a practical plan for the job to be done quickly and properly. Implement the plan and evaluate the outcome.

4. Manage time and achieve productivity

A wise person has a schedule. This is a plan for doing jobs at different times. Some jobs like sweeping the classroom have to be done every day. Other jobs, like weeding the garden or cutting grass, might be done once a week. A wise person plans times for jobs to be done so that all the tasks get done properly and quickly.

Manage time well so that you are productive in what you do in the available time. Be logical in the order of doing things. For example, to clean rooms, clean walls and dust furniture before doing the floors. Pick up rubbish, sweep and then mop. To wash clothes or kitchen equipment, start with the cleanest items and end with the dirtiest. Can you think of other examples of tasks that must be done in a logical order?

Here are some factors that influence how a person organises cleaning jobs:

- The type and size of a building, the furniture it contains and the materials from which things are made
- The number of people using a building
- Available resources: people, skills and equipment
- Amount of time available

For you to try

- Do one of the following tasks. List the times and tasks to be done. Ask yourself if time is being managed well to care for and manage your environment.
  - Make a weekly schedule for cleaning your classroom.
  - Make a work plan for yourself, from the time you wake up until you go to bed.
  - Make a work plan for your mother or father in the jobs they do each week to care for and manage the home environment.

Wise Consumer 

EFFECTIVE MONEY MANAGEMENT

A wise consumer is someone who gets value for money from the goods and services that are purchased.

Consumers are people who obtain goods and services. **Goods** are products like market foods, tin fish, rice, biscuits, clothes, cooking pots and nails.

Services are things like transport, hospitals, electricity, water supply, garbage collection and education. In modern Papua New Guinean society, kina and toea money is used to pay for the goods and services provided. Some communities still use shell money.

1. Ways of getting money

Money is an essential part of how we live in today's society. The money people earn or receive is called **income**. People in Papua New Guinea obtain their income in different ways:

- Being self-employed, selling goods such as carvings, market food or second-hand clothes
- Being self-employed, providing a service for others such as PMV transport, lawn mowing or tyre fixing
- Working as a government employee and receiving a salary such as teachers, nurses, police
- Working for an employer in the private business sector, for example, being shop workers or security officers
- Receiving rent, payments, royalties or loans from private companies or the government, for example, for land leased for housing, mines or timber projects
- From money invested, for example, in a bank or superannuation scheme

For you to try

- Survey how the parents of children in your class obtain their income.
- Invite a bank manager to explain the advantages and disadvantages of borrowing money.

Income may be regular or irregular. Government or business employees get a regular income as they receive a fixed salary each fortnight. Self-employed people are more likely to have an irregular income as they do not know how much business they will get from week to week. Artists or crafts-people earn money only when they can sell a carving, bilum or painting. It is easier to manage or budget money if the income is regular.

For some people, as soon as they receive money, they spend it. Other people plan how they will use their income.

What kind of a person will you be?

2. Reasons for managing money

There are reasons why it is good to plan how to manage your income.

- Money is spent wisely on things you really need or want. Imagine going into a shop with ten kina meaning to pay for rice and tin fish and walking out with a soccer ball. This would not be good money management.
- Some money can be saved. This money can be used for emergencies, like sickness or funerals, or expenses, like school fees or holiday travel, that you know will occur in the future.
- Managing your income is a way of showing you value the hard work that went into earning it. Imagine that your parents give you one kina each day for working around the house – gardening, looking after animals or helping to cook or to clean. You have worked hard for the money, so plan to use it wisely.
- You can provide for future improvements. Think of something you would like to improve in your life, maybe new shoes or clothes, a school bag, pen or ball. Managing your income lets you improve your life as you can afford to buy the things you would like to have.

M RECORD KEEPING

Keeping personal and business records is an important aspect of being a wise consumer.

1. Types of personal records

As a student you need to keep records like birth certificates, transcripts of academic results at school, character references and special photographs. Adults might keep records like pay slips, land titles, a will, qualification certificates, marriage certificates and references. These are personal records. It is important that they are kept in good order so that they are not lost or damaged by insects, pests or bad weather conditions.

2. Types of business records

Paper evidence of how money is used comes from business sources. A wise consumer keeps these business records in case disputes over payment arise.

A business sends a customer a bill or invoice that shows the quantity and price of goods supplied. This could be payment required for timber, electricity or rent. A person is expected to pay the amount shown on the invoice.

When the customer pays the amount on the bill or invoice, the business issues a receipt as proof of payment. This shows the amount of money paid for the goods or services received. Shops in towns give a receipt for purchases. People receive receipts when they pay telephone, water or electricity bills.

Bank statements are another example of business records. They show money coming in and out of an account over a period of time.

3. Importance of record keeping

Record keeping is important for checking on how much is earned and spent and how much is left over. It allows people to monitor their income, expenses and balance of money on a regular basis. It also provides transparency for others to see clearly whether or not a claim is true. Accurate records are necessary to show the degree of success or failure of a small business or income-generating project.

For you to try

Invite your headteacher or a business person in your community to talk to your class about the importance of keeping records on how money is used.

- Collect examples of business records. Identify the type of information provided in paperwork generated by business transactions.
- Discuss ways of storing important papers: keeping them flat to prevent damage over time; putting them in plastic folders to prevent damage from moisture; keeping them together in a folder, box or envelope so that they can be easily retrieved.
- Design and make a container or folder for storing important personal documents.

BUDGET PLANNING 

1. Composition of a budget

A budget is a plan for using money from an expected income. A budget shows the expected income and a plan for expenditure and savings. Budgets can be for a government, business, family or an individual. Here is an example of a budget for a family. This could change from week to week as income may vary and different needs arise.

A Family Budget in kina

Weekly income	150
Expenditure	
Food	60
Household needs	20
Rent	15
Fuel, firewood, kerosene	10
Transport	7
Newspapers	3
Entertainment	5
Clothes	5
Communication	5
School fees	10
Savings	10

Earn K150 a week, spend K140 a week, result = happiness
Earn K150 a week, spend K160 a week, result = unhappiness

For you to try

- Make a proposed budget for a class income-generating project. Name the project. List the things you need to buy and the cost of those items. Show the income you expect to get. Show the profit or amount you will be able to save.

2. Drawing up a budget

Steps to make a personal or family budget:

- Write how much money you get each week.
- List the things you must buy each week, for example, bus fares and food.
- List things you would like to buy if there was enough money, for example, clothes, stationery.
- Write the amount of money you can afford to pay for each item.
- Write how much you would like to save each week.
- Balance your budget. The total amount of your expenditure and savings should be the same amount as your income.

3. Why budgets are needed

A budget helps a person to use money wisely. The plan makes sure that money is allocated for necessary items. The plan allows for some money to be saved for use in the future. A budget helps people to live within their income. People who live within their income have less need to borrow money from friends or the bank. People who follow their budgets have fewer money worries.

Impulse buying is when people spend money on impulse for things that are unplanned and not budgeted for. They see something; they like it; they buy it. This is unwise as the things may not be value for money and there may not be enough money left for planned expenses.

For you to try

- Observe and ask questions to gather information on how people earn and spend money. Consider what can happen when people do not budget. Prepare a short talk on the advantages of budgeting.
- Make a personal budget. Write how much money you get or would like to get each week. Plan how to use it wisely.

4. Importance of money in modern society

Living conditions have changed in Papua New Guinea. Money is now an important factor for the quality of people's lives in modern society.

Money was not needed in traditional Papua New Guinean society. Food was grown, gathered or caught. Houses and clothes were made from bush materials. Clay or bamboo was used for cooking pots. Goods and household items were traded or bought with shell money.

Today, people need money whether they live in urban or rural areas. People like food that can be bought in shops. People like tools like bush knives and spades because they make work easier. People like clothes made of fabric because they are comfortable, colourful, available in different styles, last a long time and can be washed. People like permanent material houses because they are strong. People need to pay fees for services like health, school and transport. Young people like sports equipment and books to read for enjoyment. All these things cost money and contribute to the quality of life we want for ourselves.

For you to try

- Apply what you have learnt about effective money management, record keeping and budgeting to a practical situation.
- Start a class project to generate an income such as a garden, craft, sewing or cooking project.
- Qualities you need to be successful in business include being ready to work hard, taking responsibility for tasks to be done, being polite to customers and handling money competently. Following is an example of a project.

Project: Ice Cups (frozen cordial in plastic cups)

Budget

Expenses	
Cordial 2 × 2L @ 4.35 ea	K8.70
Plastic cups 100 for K3.00	K3.00
Electricity cost for freezing approx, K1 a day	K7.00
Total expenses	K18.70
Income	
Selling price 30 toea each × 100	K30.00
Savings or profit	K11.30

- **Investigate.** Ask questions to find out if children are interested in buying ice cups. What are the favourite colours or flavours? Identify materials and other resources needed. Is there a freezer available? Estimate costs to start the project. What price are you likely to charge? Will children be able to afford the price? Will there be a demand for your product week after week? What problems could arise?
- **Plan and design.** Plan the activities in a logical sequence. Make an action plan. Who will purchase the supplies? What colour cordial? What type of cups? How much cordial? How many cups? Who will make the cups of cordial after school each day? How will they be stacked in the freezer to become hard over night? Who will sell the product? How much money is needed to give people change? Who will count and keep the money? What records will be kept? How can everyone in the class be involved?
- **Make or produce.** Purchase the cordial and cups. Collect equipment such as a jug to add water to the cordial. Work cooperatively to mix the cordial, pour into cups and stack in a freezer. Use clean equipment. Manage time and resources effectively. Taste the cordial to see that it is the right strength and flavour. Clean up work areas and equipment.
- **Market.** Calculate the cost of production and determine a profit. Calculate a selling price. Advertise and sell the ice cups. Keep records of production costs and sales. Use a range of selling strategies. Be polite to customers. Handle money competently.
- **Evaluate.** Assess the quality of the ice cups that were made. Was the project profitable? Talk about successes, problems or failures. Make suggestions for improvements. Decide on continuing the project or the possibility of starting other projects, for example, making fruit juice ice cups or nutritious school lunch food. Write a report on the processes involved in the project. Create a financial statement for your project.

Making Things

RESOURCES USED FOR HANDCRAFTS

Many wonderful handcrafts are made and sold in markets throughout Papua New Guinea. A wide range of materials are available. These include natural resources, materials that can be reused and shop-bought new materials.

1. Natural resources

The sea, land and bush contain resources from which things can be made. These include bamboo, bark from trees, bush fibres, cane, clay, coconuts, pandanus and shells.

Bamboo is used for combs, hair clips, clothes pegs, rakes, cooking tubes, drums and furniture. Bark from trees is used for tapa cloth and clothing. Bush fibres are used for ropes, grass skirts, billums and a wide variety of baskets. Cane is used for bridges, trays, baskets, room dividing curtains and furniture. Clay is used to make cooking pots, bowls and ornaments. Coconut shells are made into bowls and other containers. Ribs of coconut leaves are made into brooms. Pandanus leaves are woven into place mats, baskets and mats. Shells are used as containers and for making necklaces.

For you to try

- Explore the characteristics of different types of locally available resources. Find and name materials that are hard or soft, rough or smooth, light or heavy, rigid or flexible, strong or not very strong and materials that sink or float.
- Investigate uses of natural resources in your locality.
-  Plan and make something from a natural resource for personal use or to sell. Make it attractive as well as useful.

Coconut bowl

You need

1 coconut
Hacksaw
Sandpaper
Piece of broken glass
Varnish and brush

You do

1. Saw coconut in half.
2. Saw off the top of one of the halves. This part is the stand for the dish.
3. Remove coconut flesh.
4. Make outside of coconut smooth using glass and then finish with sandpaper.
5. Vanish shell and let dry.
6. Place coconut bowl on the stand.

2. Used materials

Materials which can be recycled or used again include paper, rice and flour bags, cloth, old tyres, waste products and timber off-cuts.

Paper can be soaked and made into papier mâché from which containers, toys and models can be made. Rice bags and flour bags can be made into floor mats, wall hangings and carry bags. Cloths and off-cuts of fabric can be used for cleaning cloths or to sew into toys, children's clothing, wall hangings and money purses. Old tyres can be turned into pot-plant holders, garden borders and children's swings. Waste products like tins and aluminium cans are used for sawdust stoves, vases and storage containers. Timber off-cuts are used to make carvings, stools, boards for coconut scrapers, benches, shelves and minor repairs around the home.

Sawdust stove

You need

1 old paint tin
1 old fish tin
1 thick stick
sawdust

You do

1. Remove both ends of fish tin.
2. Cut hole in side of paint tin and insert cylinder of fish tin.
3. Stand a thick stick in paint tin and pack tin firmly with sawdust.
4. Remove stick to leave a tubular space that draws heat up like a chimney.

To use

Burning matter is placed in the fish tin. This lights the sawdust, which burns and heat rises. A cooking tray is placed on top of the paint tin.

Book cover

You need

Fabric, 20 cm longer and 10 cm wider than the book to be covered

You do

1. Make a narrow hem along all sides of the fabric.
2. With the wrong side of the fabric on the outside, fold fabric around ends of book and pin across corners. Remove book and backstitch to join fabric at corners.
3. Turn fabric to the right side and fit over the book.
4. Embroider your name and a small design on the front cover.

For you to try

- Identify resources in your locality that can be re-used or recycled.
- Investigate examples of recycled materials that are being used by people for purposes other than those for which they were originally intended.

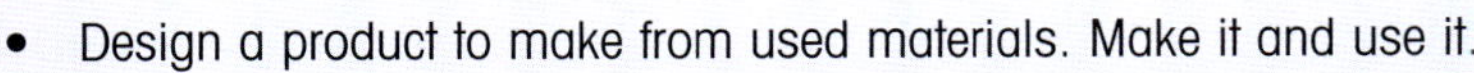

- Design a product to make from used materials. Make it and use it.

3. Imported materials

Imported materials are resources to make things. These include fabrics, metal and beads.

Fabric can be coloured using tie dye, batik and screen printing techniques. Fabric is used to make clothing like shorts, shirts, dresses and blouses. Fabric is used for cushion covers, table cloths, curtains, pillow cases and other household purposes. Metal is used to make grilling racks, barbecue plates, drum ovens, cooking trays and scrapers on coconut scrapers. Small beads are used to make armbands, ankle bands and necklaces.

Shorts

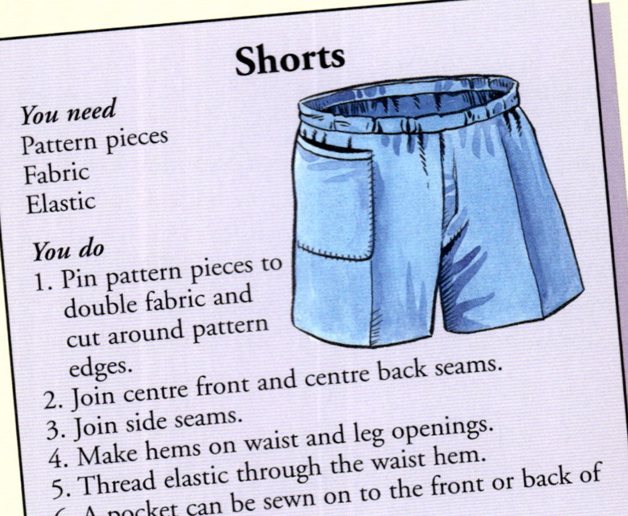

You need
Pattern pieces
Fabric
Elastic

You do

1. Pin pattern pieces to double fabric and cut around pattern edges.
2. Join centre front and centre back seams.
3. Join side seams.
4. Make hems on waist and leg openings.
5. Thread elastic through the waist hem.
6. A pocket can be sewn on to the front or back of shorts if desired.

Pillow case

You need
Fabric 150 cm × 50 cm
Tape for ties

You do

1. Fold fabric in half and join sides using backstitch or machine stitching.
2. Fold and hem the top opening.
3. Attach tape ties to fasten the pillow case over the pillow.
4. Embroider a small design on the front of the pillow case.

A TOOLS AND EQUIPMENT

1. Sewing and woodwork

Sewing

Basic tools and equipment used for sewing include sewing needles, thread, scissors, stitch unpicker, and a sewing machine.

Needles are used to pull thread in and out of fabric to make stitches. Needles with bigger eyes are easier to thread than needles with smaller eyes. Needles go rusty in damp or humid conditions and should be stored in a clean, dry place.

Reels of **thread** for sewing can be bought in many colours. Sewing thread usually matches the colour of fabric being stitched. Embroidery thread is sold in small bundles and is used for fancy stitches. Threads rot, become weak and break easily in damp or humid conditions so they should be stored in a clean, dry place.

Scissors are used for cutting. A good quality pair of scissors cuts all the way to the point of the blades. Quality scissors make long smooth cuts instead of short choppy cuts. Fabric scissors become blunt if used to cut paper or hair. Keep different pairs of scissors for different purposes. Make sure blades are kept clean and store scissors in a dry place to avoid rusting.

A **thread unpicker** is a small tool used to remove unwanted stitches. The longer point slides under the stitch and the cutting edge cuts it.

The hand-operated **sewing machine** is the most common in Papua New Guinea. It is portable and can be easily carried around. It is cheap to operate. It makes strong small stitches. Sewing by machine is faster than sewing by hand.

Woodwork

Basic woodwork tools include hammer, chisel, clamp, plane, drill and sanding block.

A **hammer** has a metal head and a wooden handle. It is used for striking nails into wood. The claw part of the head is used for pulling nails out of wood.

A **chisel** has a sharp-edged metal blade and a wooden handle. It is used for cutting wood in places where a plane cannot be used.

A **clamp** is a tool for holding wood firmly while it is drilled. A handle with a screw attachment presses the wood firmly against another section of the clamp.

A **plane** is used for smoothing the surface of a board. A plane has a handle, a cutting blade in the body part of the tool and a sole that glides over the surface of the board.

A **drill** is used to make cylindrical holes in wood. The drilling piece at the end of the tool is called a 'bit'. There are different shapes and sizes of bits used for different purposes.

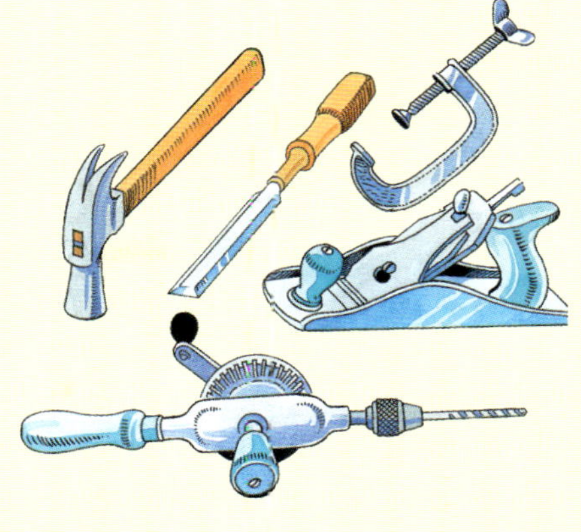

A **sanding block** is a block of wood with sandpaper wrapped around it. It is used for smoothing the surface of wood before painting or varnishing.

2. Electric tools

Some tools and equipment require electricity to work, for example, a jig saw, an electric drill and an electric sewing machine.

A **jig saw** is a tool used to saw wood into a particular shape.

An **electric drill** is used to make holes in wood where screws are to be used.

An **electric sewing machine** is operated with a foot pedal.

Care must be taken to use electric tools safely. The power point, plug and cord need to be in good condition.

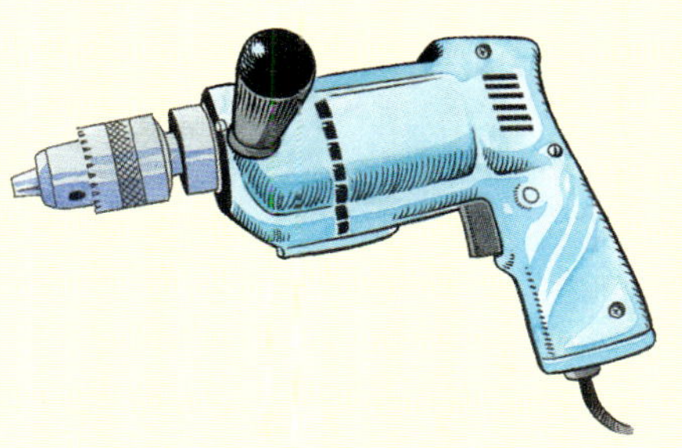

3. Care and storage of tools and equipment

Sewing and woodwork tools and equipment cost money. Building up a collection of tools and equipment costs a lot of money. Tools need to be cared for and stored well to keep them in good working condition so that they last as long as possible.

Keep tools in a special place. This could be a tool box, work bench, tool board, shelves or a cupboard. The storage place should be clean and dry as dampness causes tools to rust. Dirt and dust cause tools to not work smoothly.

Identify a suitable place for sewing and woodwork tools to be well and securely stored. Have a place for everything and keep everything in its place when it is not in use. Label shelves. Keep a stockbook to record the number and condition of tools and equipment. Create an effective borrowing system to make sure that tools are returned after use.

4. Maintenance and repair

Tools and equipment need to be maintained and repaired to prolong their usefulness.

Tools and equipment are maintained through correct handling and storage. They should only be used for the purpose for which they were meant to be used.

Sewing machines should be dusted and oiled and kept covered when not in use.

Equipment should be repaired when it becomes damaged so that it can be used safely.

For you to try

- Observe and experiment with how to handle tools and equipment properly and safely.
- Name tools, their parts and their uses.
- Observe and experiment to find how to clean and oil a sewing machine.
- Collect tools and equipment that are damaged and carry out appropriate repairs to put them in good working order.

A TYPES OF HANDCRAFTS

Many types of handcrafts are made in Papua New Guinea. They include sewing, weaving, mats, photo frames, baskets, trays, bilums, soft toys, hangers, flowers, table decorations, designs, flower vases, carving, painting, jewellery and traditional **bilas**.

Wooden pencil case

You need

2 pieces plywood 210 mm × 50 mm × 10 mm for sides
2 pieces plywood 60 mm × 50 mm × 10 mm for ends
1 piece plywood 210 mm × 80 mm × 3 mm plywood for base
1 piece plywood 210 mm × 80 mm × 3 mm plywood for lid
Sandpaper
Small screws and nails
Varnish and glue

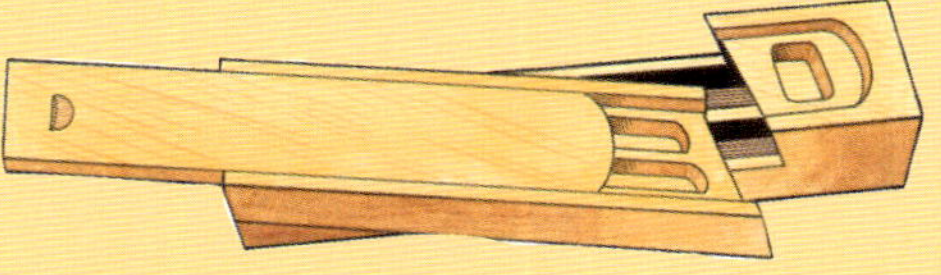

You do

1. Prepare pieces of plywood needed to make the pencil case.
2. Saw the lid as shown in the top view diagram.
3. Sandpaper surfaces and edges before gluing.
4. Glue and nail pieces together.
5. Drill hole and counter-sink the screw on the lid.
6. Finish by varnishing or painting.

Small drawstring bag

You need

Fabric 30 cm × 45 cm
String 65 cm

You do

1. Fold fabric in half.
2. Pin, tack and backstitch two sides together.
3. Fold a 2 cm hem on the top edge.
4. Make a small buttonhole on the outside of the hem.
5. Thread string through the hem and tie ends.
6. Embroider your name and a design on the bag.

Wall hanging with pockets

You need

Fabric 50 cm × 30 cm
2 pieces of fabric for pockets 20 cm × 15 cm
Small stick and string for hanging

You do

1. Make 1 cm hems on long side of fabric.
2. Make a 5 cm hem on top edge for stick to thread through.
3. Make 1 cm hem on top edge of pockets.
4. Fold under 1 cm on other edges of pockets and pin into position on the wall hanging. Hemstitch or machine along edges.

For you to try

- Explore ways to use available resources to make attractive and useful handcrafts to sell.
- Produce an action plan.
- Make a handcraft. Decide upon a suitable selling price. Advertise and market it.
- Evaluate the result of your handcraft project.

USING AVAILABLE RESOURCES

It is important to use and value available resources to make things.

- Available resources are cheap and easy to get.
- They preserve Papua New Guinea's culture and give a national identity to the products that are made.
- Products are original and appeal to both local people and tourists.
- Local resources stimulate creativity in people, as they adapt them for different designs and uses.
- People take pride in making home-made products.
- Products bring out the beauty in the materials being used.

For you to try

- Bring a handcraft item to school. Describe how it is made and its usefulness. Describe the characteristics of the materials and explain why they are suited to the purpose of the handcraft. Consider structure, texture and colour of the materials for specific purposes. Explain the advantages of using available resources.

Community Development

Chapter Summary

In this chapter you will have an opportunity to:

- ✔ **Investigate and identify common features of your local community and how these features benefit or influence the people living there.**
- ✔ **Learn about the importance of effective communication and investigate the different mediums of communication.**
- ✔ **Practise the skills of effective communication.**
- ✔ **Research and plan suitable projects to meet the needs of your school and local community.**

Syllabus references

Strand: Community Development

Substrand 1: Knowing Communities

Outcome: 6.3.1 Identify the common features of the local community and make suggestions as to how these may influence or benefit community life.

Substrand 2: Communication

Outcome: 6.3.2 Investigate and practise practical skills for effective communication

Substrand 3: Community Projects

Outcome: 6.3.3 Participate in projects to meet identified needs and opportunities for improvement within the school

Knowing Your Community

People have always lived in groups to support each other in meeting their basic needs for survival. One of the most important groups that people belong to is a family group. Whenever people live together they form communities. Some communities are large; others are small in size.

A small community

Simbu village near Chuave

A large community

Goroka and the Asaro Valley

Communities can be found in many different places. Some are set up in and around urban areas like towns and cities. Other communities are located in rural areas, in the villages. In Papua New Guinea, about 85 per cent of the total population of five million people live in rural communities and only 15 per cent live and work in urban communities.

People in rural communities spend a lot of their time gardening, looking after animals and growing cash crops such as coffee, vanilla, cocoa and coconuts to make copra. Sometimes they go hunting and fishing. It is through such activities that community members meet their basic needs for survival. Rural communities often do not have well-developed government services such as health care, transport and education services.

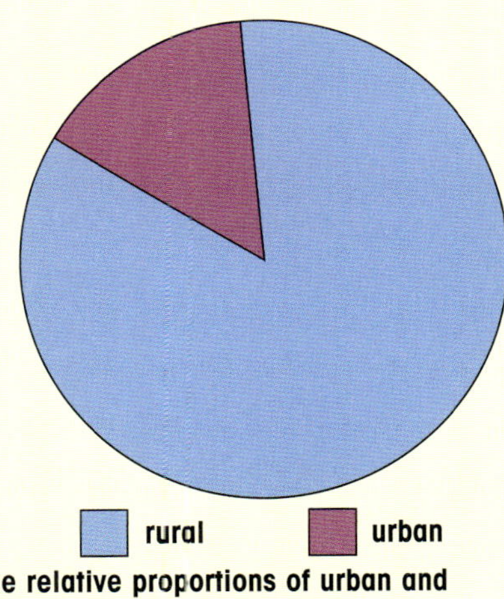

The relative proportions of urban and rural populations in Papua New Guinea

PD COMMON FEATURES OF A COMMUNITY

What is a community?

A community is made up of a group of people living in the same area who work together to meet their needs and wants.

You may come from a village, town or city. Within each of these communities you belong to even smaller communities; the obvious one is your family. In Papua New Guinea communities are often made up of family groups, which form the village. The family is the most important unit in the community. Your extended family is made up of your aunties, uncles, cousins, grandparents and other distant family members who form the base of the community.

Many of these communities share and speak a common language. This helps them to identify with other members of the community.

For you to try

- Go out into the community and listen and ask questions of a community leader such as church pastor, teacher or local counsellor, about the community.

1. Basic services

These are the many things that make up the community and affect the lives of people in the community. Services such as bridges, roads, schools, aid posts and church are there for the benefit of people. The government, churches, overseas donors and other NGO groups help develop services to meet community needs.

Some features of a community are improvements made by people, such as roads, bridges, wharves, airstrips, schools, buildings, water supply, health centre and playing fields. Other things, like weather, landforms, rivers, mountains, land and the coast, are not directly influenced by people. Both the natural and human-made environments benefit and influence communities in Papua New Guinea.

The *Yumi Yet* bridge program was started by the government and work is carried out by a private company. This program is responding to the need for better and stronger bridges, especially in the rural communities of Papua New Guinea. It is making travel safer and easier.

Here is a map of a village community located in Papua New Guinea. A community resource map shows the common features that make up the community.

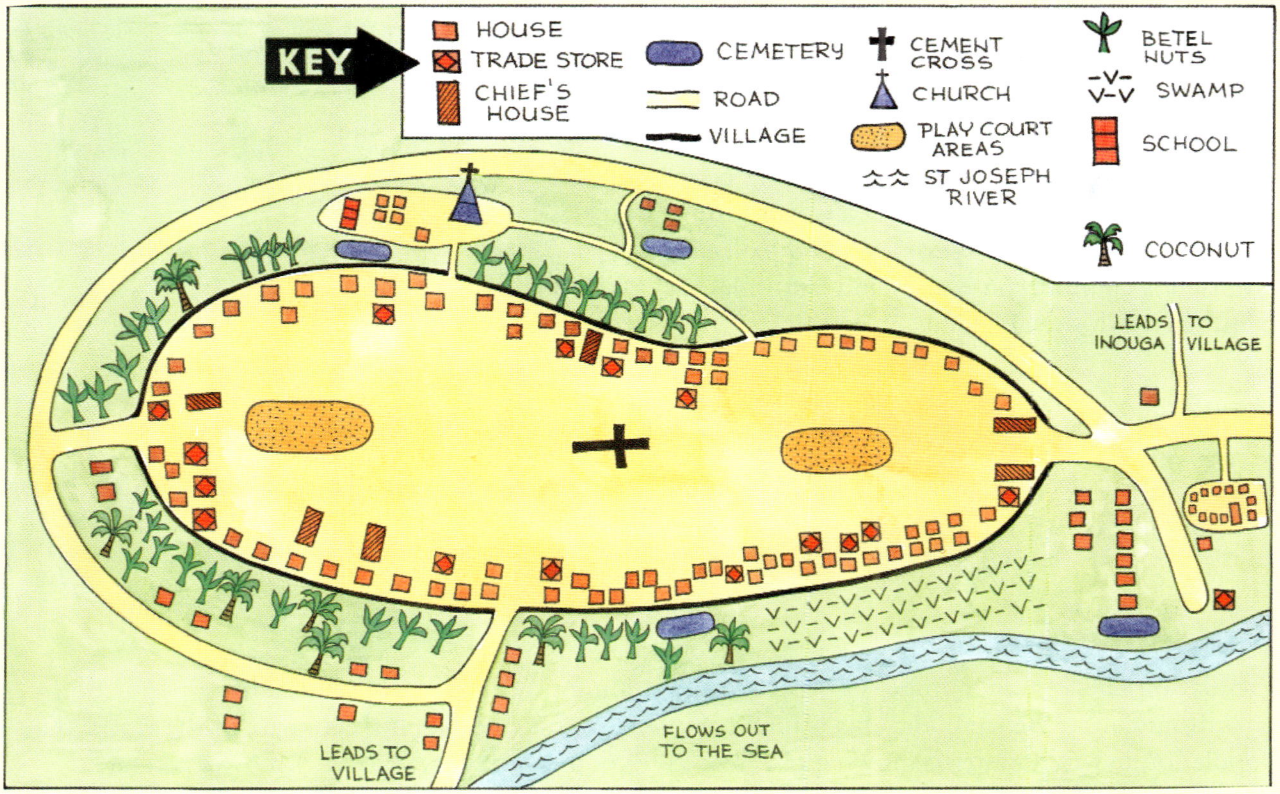

KEY
HOUSE
TRADE STORE
CHIEF'S HOUSE
CEMETERY
ROAD
VILLAGE
CEMENT CROSS
CHURCH
PLAY COURT AREAS
ST JOSEPH RIVER
BETEL NUTS
SWAMP
SCHOOL
COCONUT
LEADS TO INOUGA VILLAGE
LEADS TO VILLAGE
FLOWS OUT TO THE SEA

Village Map of Rarai

For you to try

- Walk around your local village or town and record key features. Use symbols, pictures and key words to show them on your map.

2. Geographic locations SS

Location refers to the placing of a community. There are things in the environment that are useful in making life easier and enjoyable for the people who live in communities. These factors determine where a community is located. For example, a community built close to a river provides water for washing and drinking. Travel is easier using canoes and food resources can be harvested from the river. River valleys are also rich in fertile soil for growing food crops.

There are many different landform features in Papua New Guinea. We have the lowland coastal areas, islands and atolls, coastal plains, river valleys, mountains and plateaus. People chose where they would like to settle for many different reasons such as safety, resources available and easy access to services.

A community on the coast may have plenty of seafood to eat and sell to make money. Children enjoy swimming in the sea and use the beach to play on. Plains and river valleys make fertile soils for gardens and produce can be sold at markets or eaten. A community located on a mountain can be safer from attack by enemy tribes.

However, natural disasters can endanger people's lives. On the coast we have tidal waves and tsunamis; plains and river valleys experience flood; volcano eruption is common on islands and landslides are common up in the highlands.

Fly River

For you to try

- Walk around your school area and observe the land features on which your school is built. List advantages and disadvantages of your school's location.

SS 3. Populations

Population refers to the number of people in the community. Papua New Guinea has a population growth rate of 2.3 per cent and this is high compared to some developing countries in the world. The population in this country is growing fast. The 2000 population census (count) showed a total population of 5.2 million.

However, the population is not evenly spread throughout Papua New Guinea. Some parts have more people than others. In the Simbu Province as well as the other highlands provinces, there are more people than in places like the Gulf and Western Provinces. Improved law and order and farming in the highlands helped the population to grow. There are not as many people in the lowland plains, like the Sepik, Gulf and Western Provinces, because many people fall victim to sickness, such as malaria. Also, few people are found in the lowland plains because of the thick forests and swamps.

The number of people living in towns and cities is growing very fast. People move into towns because it is easier to reach basic services like schools and hospitals. Others have run away from sorceries and the hard life in villages.

People are important resources because they have the skills and knowledge to turn natural resources into something useful or they can use resources to benefit themselves and others.

For you to try

- A Conduct a census of your school, village or local community.
- Represent your information as a graph.

4. Culture PD

Culture is the way of life. Each community has its own customs and traditions that distinguish it from other communities. This helps people identify with their own community members.

The type of food eaten, different styles of houses, clothes, courting practices and ways of behaving make up our culture. People in the highlands build their houses on the ground with no windows because it is cold and to keep the heat inside. Coastal people build their homes on stilt posts and windows keep them cool.

To welcome visitors, the people of the Western Highlands Province cut sugar cane from their gardens to give to them; the Manam Islanders in Madang Province welcome their visitors with kulau drinks and offer betel nut to chew. Dancing to the bamboo flutes in North Solomons is different from Mekeos of Central Province with colourful headdresses and following kundu drums.

The different cultural practices are learnt by young children through living and practising cultural activities under the watchful eyes of the elders. The elders are the keepers of community traditions, legends and histories and they pass on this knowledge to younger generations. All people have a responsibility to look after and educate children and to care for older people.

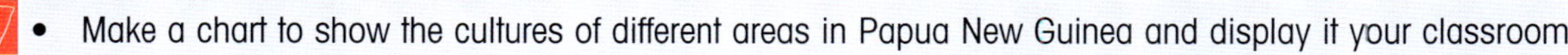

For you to try

- Plan a cultural day for your school.
- Make a traditional item from your local area.
-  Make a chart to show the cultures of different areas in Papua New Guinea and display it your classroom.

5. Climate SS

Climate is the average weather of an area and much of Papua New Guinea is hot and wet all year round. This is because Papua New Guinea is located in a tropical region along the equator. However, within Papua New Guinea there are differences in climate. Coastal lowland areas are quite hot while the highlands region is cold.

The climate of an area determines the type of plants that grow there; the crops that people grow in their gardens; the style of houses that are built; and the activities people do throughout the year.

The two main seasons are the wet and the dry. People plan their activities based on these two seasons. Gardens are prepared during the dry season – bush is cleared and burned. Planting is done just before it rains and, following that, weeding takes place until the food is ready for harvesting. On the coast, people go out fishing at night when the sky is clear and the moon is bright, usually during the dry season.

Sometimes we have too much rain during wet seasons and this floods lowlands and riverbanks in places like Sepik and Ramu. Floods wash away food gardens, roads and bridges. Domesticated animals and people drown. Floods can also cause landslides. At other times we have extended periods of dry. This is called **drought**. Droughts can kill plants and animals due to shortage of water. People can also suffer and die from food shortage and lack of drinking water.

For you to try

- Make a fan from local materials.
- Make neckwarmers or small heater to generate heat in cold regions.
- Investigate how the climate of your local area influences people and the type of activities they do.

SS 6. Natural resources

Natural resources are useful things found in the environment that are created by nature and not made or influenced by people. Natural resources are things like forests, mountains, rivers, sea, oil, minerals and the land.

Some of the natural resources found in our environment can be renewed after they have been used. Water, soil, air and trees are known as **renewable** resources. Other resources cannot be renewed when they are used up, for example, minerals, oil, gases and fuels. These are called **non-renewable** resources.

People have to be careful in the use of natural resources so that we continue to benefit from them now and in the future. This is what our Constitution says about the use of natural resources and this is known as **conservation**. Conservation is essential so that future generations will not miss out on the benefits of the natural resources found in our environment.

Logs at Hoskins ready for export

For you to try

- Walk around your village community and observe all natural resources.
- Make a list of these natural resources.
- Classify the items on your list under the headings 'Renewable' and 'Non-renewable'.

7. Community members PD

A community is made up of people of different sexes, ages and skill levels. In a community you will find children, young people, women, men, community elders and community leaders. Some important and key members of the community are government officials such as health workers, teachers and agriculture workers. We also have church and community elders who make sure that the social and spiritual needs of people are cared for.

Members of a community have particular parts (or roles) to play. People become good at doing certain activities and then share their skills and ideas to benefit the whole community. Elders are the keepers of community traditions, legends and histories, and they pass this knowledge on to younger generations. Government officers help people in the community by working to provide important government services like health and education.

For you to try

- Interview a community member to tell your class about his/her role in your community and ask questions about what they do.

- Use the information told to the class by the community member and complete the table below. Copy and complete this table in your exercise book.

Community member	Role played
Ward councillor	
Community elder	
Teacher	
Youth leader	
Pastor/priest	

RESOURCES THAT BENEFIT A COMMUNITY

Everywhere, people can benefit from the unique features of the place where they live.

- School gives children the opportunity to be educated.
- A Health Centre enables you to get treatment quickly when you need it.
- PMV services allow you to travel to other places if you have the money to pay the fare.
- Roads give people access to other places to sell products or services.
- A nearby sawmill allows you to get timber off-cuts and sawdust to use at home.
- Trees provide firewood.
- A trade store sells goods like kerosene, matches, soap, tinned fish and rice.
- A church provides the opportunity to hear the word of God and develop spiritually.
- Living on the coast allows families easy access to fish and other seafood.
- Living in the highlands allows people to enjoy cool weather.
- Fertile soil enables people to grow healthy crops.
- Connection to an electric power supply means people can have good lighting and use electrical appliances.
- A water well gives families access to a clean water supply for cooking and drinking.
- Employment opportunities allow some people to earn a regular salary.
- Cash crops enable a family to earn cash income.
- Living in or near a town gives families access to big businesses in the province and government headquarters.

Picking coffee beans

For you to try

- Go on an excursion into your village or town and observe features that may benefit the community.
- List the features in a table under the headings 'Feature' and 'Benefit'.
- What other features would you like to see that could benefit your community?

Read this story about Michael from Riwo village in Madang.

Michael comes from a coastal village called Riwo, about eight kilometres along the north-coast highway, fifteen minutes drive from Madang town. The road connecting Riwo village and the main town is sealed with coal tar and many PMV trucks and buses use the road to transport people, cocoa, copra and garden foods to the main centre. They also carry trade store cargo and other factory-made goods back to the village.

Like all other coastal villages in Papua New Guinea, one of the important natural resources for the Riwo community is the sea. The villagers use the sea for washing and picnics but drinking water is collected from tanks and the bore-well close to the village. Fish and other seafood gathered from the sea is either eaten or sold at local markets for some cash income. Michael's people are farmers and they grow cocoa and coconut to make copra to sell. The land is also good for growing food crops to feed families. They grow bananas, yams, kaukau, aibika, pitpit, and fruits like pawpaw, mango and guava. Any left-over garden produce is taken to the markets for sale.

The sale of seafood, cash crops and surplus garden food gives people some cash income to pay for school fees, meet hospital costs and buy other trade store goods such as rice, kerosene, soap and tinned fish. These enable people to enjoy a good lifestyle.

Other people from the Riwo community work for private companies or the government at the provincial head office in town. They travel to their place of work and back using the PMV system provided by local operators. Michael works in town for a government training college and has to travel every day to and from work, using the PMV system provided.

At the Riwo village community, there are many buildings. Some use timber from nearby bushes for the posts. The roof, walls and flooring are made of sago thatches and coconut fronds which are plentiful. A good number of buildings are made of modern building materials and they are connected to electricity supplies from Yonki hydropower. Most buildings are for living but some are used for church, aid post, copra drier shed and other purposes. The main building at Riwo is the Catholic Church where most of the 600 men, women and children belonging to the Catholic faith attend Sunday services. Michael is a leading member of the Catholic Church and he is involved in church activities to promote the spiritual growth of community members.

The children at the village community attend school at Ambarina Primary School, which is about two kilometres from the village. Others attend Sagalau Primary School, which is also a very long way from the village. The children travel back and forth by bicycle or PMV trucks. Other times they walk the long distance to get to their schools.

For you to try

- List resources that feature in Michael's village at Riwo.
- Pick out things in Riwo village that are similar and things that are different to your village community. Draw a table to show these similarities and differences.

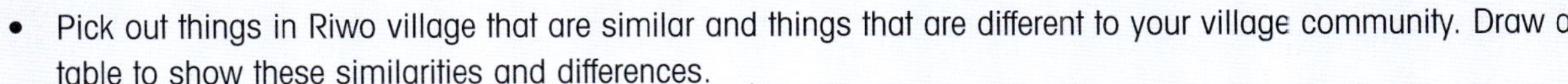

1. Services

Easy access to services provided by the government, community-based organisations and non-government organisations is very important. Community services such as schools, roads, public transport, electricity, drinking water and industry make life a lot easier and more comfortable.

In Papua New Guinea most people do not have access to basic services. These services cost a lot of money to put in place and the community needs to look after them, if they are lucky enough to have access to these services. In this way communities can continue to enjoy a good life over many years.

Electricity in your home helps you to study at night, cook meals using electric stoves, watch the local EM TV news, keep your food fresh in a freezer and listen to music from your hi-fi stereo. Public transport helps you to travel to and from school, brings your garden foods to market, transports copra or coffee to the factory and brings store goods to your community. All these services require good roads. A supply of fresh and clean water to your community saves walking long distances and leads to healthy living.

Steep mountains, fast flowing rivers and thick forests can make it difficult for important government services to reach people. This makes it difficult for people to access road systems, schools and health services. Difficulty accessing basic services is a reason people move from their village communities.

For you to try

- Carry out community service work around your school by picking up rubbish, cutting your lawn, or helping cut grass along the road or clean the community market grounds.
- A Run an awareness campaign on looking after services provided by the government and other organisations. Focus on a service related to your community such as community notice boards, road signs, school buildings, health centres and water supply pipes. Prepare charts and other aids to better explain your points.

2. Locations (SS)

Benefits of living near a river

- The water from the river is used for drinking, washing and fishing.
- The river is also used for watering gardens and to travel from one place to another using canoes. Using canoes to travel on the river, people can visit their gardens to collect food, go fishing or visit places where they make and collect sago.
- Riverbanks are rich in fertile soil left behind when the river floods on the flat land close to the banks. This is good for growing food crops.

Rivers are important resources that benefit communities. However, a river can be a problem for people who wish to travel on land. People can only cross rivers where they are not deep or where there is a bridge.

During rainy seasons rivers can be a problem when there is a flood. It can wash away food gardens, animals, roads, bridges and good garden soil.

Benefits of living in the highlands

- For communities in the highlands, the natural resources of the forests meet all needs. People gather fruit and nuts to eat, hunt animals, collect wood for fires, and use timber from trees to build fences and houses.
- People use the land for subsistence farming and a great variety of food can be grown. This means that communities can rely on growing all their food.
- People enjoy the cool weather experienced up in the mountains.

Mt Elimbari in Chuave, Simbu Province

Benefits of living on the coast

- For communities on the coast, the sea is an important resource. Many groups of people live on the coast.
- They collect most of their food from the sea.
- Their houses are usually built close to the shore to protect their canoes.
- They choose places that are protected from strong winds and large waves so that houses are not destroyed. Large bays and inlets are safe and many village communities and some coastal towns can be found in these places.
- People are influenced by the changing tides to do fishing and to collect seafood.
- A traditional calendar is used to plan for long sea journeys to trade and exchange goods. They use specially built seagoing transports.

Waiori village on Marshall Lagoon

For you to try

- Use information on benefits to the community and fill in the table below.

Location	Benefits	Disadvantages
Living near a river	1	1
	2	2
Living in the highlands	1	1
	2	2
Living near a coast	1	1
	2	2

- Go on an excursion around your local area and observe benefits and disadvantages. Add your information to the table.

3. Populations 

The number of people in both rural and urban communities is increasing because of improved health services and better nutrition. Also, a lot of people are changing locations, especially from rural locations to urban areas. That is causing an increase in urban populations. Some problems related to an increasing population are shortage of land, overcrowding and food shortage.

There are not enough jobs for everyone in the country, and this means that there is unemployment. Unemployment is a problem in town settlements where money is needed to pay for food, clothing, education and health services. Unemployed youth get involved in criminal activities such as pick-pocketing, stealing, prostitution and armed robbery to get what they need to survive.

In the highlands, land shortage is a big problem due to the increase in population. Many tribal fights result from arguments over who owns land. The landform is mountainous and good garden land is in short supply when overcrowded with people. Land is passed on from father to son. People who do not want to argue over land move on to other places.

People who move to other places make homes close to towns and cities on either traditional or government land without proper permission being given. Those who settle on such land are called squatter settlers. This is a growing problem in Papua New Guinea.

For you to try

- Count the number of family members of your classmates to work out the total population of people related to your class. Present as a graph or table.
- What are the advantages and disadvantages of:
  - having a big family?
  - having a small family number?

4. Natural resources

Many groups of people live on the coast. They collect most of their food from the sea. Seafood gathered can be used to feed families or be sold in local markets. The sea can also be an attractive area for tourists to visit. Sand from the beach can be used for buildings and making footpaths.

The natural resources of the forests meet needs for communities throughout Papua New Guinea. The type of forest found in our country is tropical rainforest and there is a great variety of plant and animal life. People gather fruit and nuts to eat, collect wood for fires and use timber from trees to build fences and houses. Plants in the forest provide useful medicines. Also many wild animals are hunted for food.

The water from rivers is used for drinking, washing and fishing.

The river is also used for watering gardens and to travel from one place to another using canoes. Papua New Guinea is rich with natural resources. Some of these natural resources, like rivers and forests, are renewable and others, like minerals and oil, are non-renewable.

Natural resources can benefit local communities and the people of Papua New Guinea. Foreign companies working in Papua New Guinea take out natural resources from the ground, sea and forests. These are sold overseas where they are made into finished goods. We earn money from the export of natural resources.

For you to try

- Invite a local community leader or elder from your village or town to explain what is the most important natural resource for that area.
- Walk around your community or school and identify natural resources.
- Write a report to show how the resources are used in your community.

5. Climate and seasons SS

Climate and the change in seasons can assist or hinder people in the community with their everyday activities. Some of the activities that can be affected are:

- Agricultural activities, like cash crops. These can be affected if there is too much or too little rain. Drought can badly affect farmers of tree crops like coffee, copra and cocoa. Floods can destroy vanilla gardens and coconut trees.
- Food garden preparations and planting. Most communities in Papua New Guinea use the dry season to clear bush for cultivation of soil and to plant crops before the start of the wet season which helps crops to grow well. The food calendar starts with planting during the months of December through to March, during the wet season. Harvest occurs from September through to November, during the dry season.
- Sports, feasts, dances and important ceremonies. These are planned to take place when it is fine, during the dry season.
- Food production in different communities. Kaukau grows well up in the highlands where it is cold and sago palms, taro and yams grow well in coastal plains where it is hot and humid. These are the main foods for communities on the coast and highlands of Papua New Guinea. No food is better or worse than the other because it is cultural and relates to the identity of a community.

Traditional calendars are used to plan the community's economic and social activities. Elders in the community use their experiences and traditional knowledge to plan activities that go in line with seasonal changes. These include important activities such as fishing, gardening, feasting, dancing and hunting.

For you to try

- Invite an elder to explain a traditional calendar used in your area for activities such as:
  hunting
  fishing
  clearing bush
  weeding gardens
  planting crops
- A Draw up a traditional calendar based on this information.

Communication

Communication is a process which involves sending, receiving and interpreting messages. It is important for people to share information. They need to send and receive information. This process may take place internally (within ourselves), between two people or a small group, or publicly before a large number of people.

Communication is a two-way process that involves both the sender and the receiver. Messages are sent by the sender, received and interpreted by the receiver, and then returned to the sender.

There are three elements in the communication process.

1 The **source** is where the message comes from. Messages can come from one person, or a group of people or through an organisation like the media – newspapers, radio stations or television.

2 The **message** is the information that is sent. It can be sent in any form – ink on paper, through sound waves in the air, or any other signals you can interpret and understand.

3 The **destination** is the place or person the message is sent to and received. It can be sent to one person or a group of people listening, watching or reading.

IMPORTANCE OF EFFECTIVE COMMUNICATION

Communication is very important in our daily lives. We need to communicate within our families and communities and with the rest of the world. Communication binds people together. Without communication, there would be no cooperation, no interaction and no organisation. There would be total disorder in our families and communities and throughout the world.

We need effective communication to:

- **Keep the community informed.** This means people in the community need to share information through conversations, announcements and meetings – community meetings, community notice boards, public awareness and so on.
- **Develop relationships** with people by meeting people socially or for business or by sharing similar ambitions and interests.
- **Network within groups** for the benefit of individuals and also the community group, be it social or economic.

For you to try

- Interview your school principal to discover how your school informs your local community of school news (school fees, assessment results, parents' day, school fete day . . .)
- Present your findings as a written report.

MEDIUMS OF COMMUNICATION

In Papua New Guinea, people still use traditional as well as modern forms of communication. In the past people only needed news about their own community, but today people want news about a much larger community and this includes world news. This information comes through the media.

Traditional mediums of communication vary from one society to another, and only people from the same society may be able to interpret and understand the message sent. Often traditional communication is slow and it reaches few people. However, with modern methods of communication, all people in the country can receive messages quickly and understand the messages. Modern forms of communication include radio, telephone, newspapers, television, computer letters, fax and Internet.

Modern mediums of communication

1. Newspaper

A newspaper is a number of large sheets of folded paper on which news articles, advertisements and other information are printed. In Papua New Guinea, we have two main newspapers printed every day from Monday to Friday. They are the *National* newspaper and the *Post-Courier* newspaper.

For you to try

- In small groups review a newspaper and identify the main sections.
- Design an advertisement promoting your school.
- Write a story about yourself and your school and send it to the *National* newspaper to be printed in the children's section called 'Young Life'.

Note: the children's section is printed every Wednesday.

2. Letters

When you write a letter, you write a message on paper and send it to someone, usually through the post. You put the written message into an envelope and post it through the post office. It is also called **mail**.

Useful hints when posting your letter:

- Write your address correctly in the middle of the envelope.
- Affix the correct value stamp onto the right-hand corner of the envelope.
- Drop your letter into the correct box. (Local, PNG, Overseas.)
- Post your letter early in the day.
- Check your mail box for incoming mail.

For you to try

- If your school is close to a town, visit a post office and observe how it is run.
- Write and send a friendly letter to one of your relatives.

3. Fax

A fax is a message or letter sent through a fax machine. An exact copy is produced in another fax machine in another place. The message is received in the same minute. It is a fast way of sending written information. People can pay for fax machines to be connected into their houses and offices. If you want to send a fax message, you can go to the post office to send your message. Normally, you will pay for each page you send through the fax machine.

How to send fax message

- Insert the written message/paper correctly into the fax machine.
- Dial the fax number of the person or office you want to send your message to.
- Wait and the fax machine will tell you whether the message you sent has gone through to the other fax machine.

For you to try

- Find the fax number for the *National* newspaper and fax a letter to the Editor explaining some special activities that you are doing at school.
- Send a fax to the newspaper to invite a reporter to visit your school to write a story about your school.

4. Email

Email works like a fax. When you want to send an email message you use a computer. Type the message into the computer and send the message through the computer to the email address of the person you want to contact. The message will be sent to another computer in another place. It is the safest and fastest way of sending and receiving information.

How to send an email

- Switch on your computer.
- Open up the email program.
- Follow the instructions given on the computer screen to send your email message.
- Type your message on the computer.
- Send your message to the email address of the person you want to contact.
- Check your 'in box' for a reply to your email message.

For you to try

- If your school has a computer, set up a penfriend club by emailing class-mates' names to twenty Australian schools. Ask your teacher to obtain the email addresses for you.

5. Telephone

A telephone is an electrical system used to talk to someone in another place by dialling a number on a piece of equipment and speaking into it. A list of names, addresses and telephone numbers of people with telephones is contained in a telephone book. People can pay for telephones to be connected into their houses and offices. A telephone box is a small shelter in the street which contains a public telephone. Some telephones are connected for the public to use. This place is called a telephone booth. Usually, you have to pay to use the telephone.

How to use a public telephone

- Pick up the mouthpiece and hold it close to your ear and mouth.
- Insert coins into the telephone (if the telephone accepts coins) or insert a telephone card (if the telephone accepts telephone cards).
- Wait for the dial tone and dial the telephone number of the person you want to contact.
- When the person answers your call, you can begin talking. If the telephone line is engaged, then you will hear a pipping sound. Wait a while and dial the number again.
- After talking, hang the mouthpiece back on to the telephone.
- Do not forget to collect your unused coins – they drop into the coin box – or pull out your telephone card.

How to make IDD calls

- Dial the international access code '05'.
- Dial the country code.
- Dial the area code.
- Dial the telephone number.

Types of telephone calls

- **Local Calls** are calls you make within the same area or town. They are not timed and you are charged a single 30t fee.
- **STD Calls** are calls you make within the country. The call is timed and you are charged 30t for every minute.
- **IDD Calls** are calls you make to other countries. The calls are timed and the charge is high.

For you to try

- In pairs, collect a long string rope and two empty tins of fish or Milo. Tie the empty tins to each end of the rope. Act out a drama, sending messages as if through a real telephone.

6. Television

Television is a system of sending pictures and sounds by radio waves so that people can watch them on a television set at a distance from the point of origin. You can watch commercial advertisements, movies, national and world news and many other programs.

In Papua New Guinea, our television station is called EMTV.

For you to try

- Invite a TV reporter to your school to film special activities.
- List the different types of programs shown on EMTV.
- What TV programs do you like to watch and why do you like these programs?
- Discuss the advantages and disadvantages of television.

7. Radio

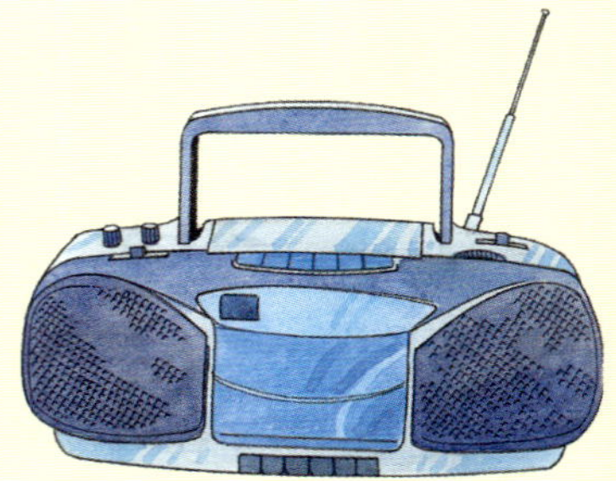

Radio is a system of sending and receiving sound messages broadcast on airwaves through radio stations. There are many different radio stations each with a different channel number. You can select and tune into the frequency number of the radio station and listen. There are provincial radio stations and national radio stations.

For you to try

- If your school is close to a radio station go on an excursion to find out how a radio station works.
- What is your provincial or local radio station called?
- List the radio programs you like listening to on your local radio stations.

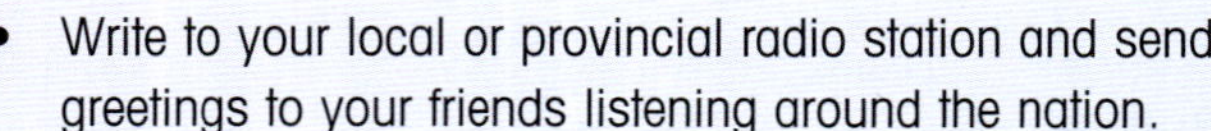

- Write to your local or provincial radio station and send greetings to your friends listening around the nation.

EFFECTIVE COMMUNICATION SKILLS 

Listening to people who talk to you, speaking to people, reading information in newspapers and books and writing messages are very important communication skills. All these skills – speaking, listening, reading and writing – are interconnected and rarely used in isolation.

1. Listening and speaking

Listening and speaking are interrelated. We listen to people while they are speaking to us and respond to them by speaking. This is an oral form of communication. We use this method in our daily life to receive and share information with our families, at work and within the wider community.

For effective **listening skills:**

- Do not speak when you are listening to the speaker.
- Do not interrupt the speaker.
- Keep eye contact with the speaker.
- Ask questions of the speaker at the end of the conversation if the message/information is unclear.

For effective **speaking skills:**

- Speak clearly.
- Make sure your message is well structured.
- Be prepared to answer questions to back your message.

Interviewing is one method of oral communication. To conduct an interview you must:

- have a topic on which you wish to gain information
- prepare questions that will give you information about the topic you are researching
- prepare a recording sheet so that people's responses can be recorded
- select people to be interviewed and explain the purpose of the interview
- carry out the interviews
- fill in the response sheets
- summarise your data and write up your findings

For you to try

- Identify an urgent need in your school. In pairs prepare interview questions to find out what is urgently required in the school. Interview at least twenty students from different grades and present an oral presentation to your class.
- Using the above procedure, interview a classmate about his/her family history.

2. Reading and writing

We can also communicate with each other using written symbols, signs and words. These are written down and we need to read and interpret to understand them. We can then respond to this written form of communication by replying in written symbols, signs and words.

Today, we use modern methods of communication. People can read all sorts of writings through newspapers, books, pamphlets and computers.

There are many forms of written expression depending on its purpose. We write

- **personal letters** to friends telling them personal news
- **formal letters** when we apply for a job
- **specific instructions** for making things, cooking, giving directions and so on
- **information reports**
- **information recording templates** such as surveys, questionnaires and summaries

For you to try

- Walk around your local community and observe signs and symbols. List signs and symbols and say what they represent.
- Conduct a class discussion on *Why do we have newspapers?*
- A Prepare a written summary of what you discover and present a summary to the class. Is your summary different from that of your classmates?

3. Collecting data and information

Collecting data and information from people is important when making important decisions. The following methods are used for collecting data and information:

- **Surveys.** An interviewer talks with many people to gauge their views, opinions, ideas and suggestions on an issue. The interviewer then compiles responses and writes a summary or report based on the responses.
- **Questionnaires.** Questions are written on a sheet of paper, leaving spaces for people to write their responses. Responses are tallied and this helps to draw conclusions and make recommendations.
- **Computers programs**. Computers are effective tools for data collection, data recording and analysis . Once information has been fed into a computer program, data can be analysed very quickly.

For you to try

- Write down interview quesitons to find out the type of jobs students hope to get when they finish their education.
-  Interview ten students. Prepare a written summary and present your findings to the class.

4. Using the computer

We now live in a computer age. Computers are used in industry, offices, schools and homes. Although computers are expensive to buy, more and more people have discovered that they are an effective mode of communication as well as a reliable tool for storing and relaying information quickly and effectively.

The basic computer programs for beginners are Microsoft Word and Microsoft Excel.

For you to try

- If you have a computer in your school or home, practise your typing skills so that you can use the Microsoft Word program.
- If you have the Microsoft Word program at school, use it to type up a school investigation.

Community Projects

COMMUNITY NEEDS AND OPPORTUNITIES

SS 1. Needs

People on Earth need food, air, sunlight and water to survive. We need safe drinking water to drink and to cook. We need nutritious food to eat and a balanced diet daily. We need clean air to breathe. We need safe shelter to live. We need a healthy environment to live in. These are the essentials for life. While some of these things are provided by nature, others we need to provide for ourselves.

The air we breath, sunlight and the water we drink are provided from nature. We get them for free, but we must protect them from pollution so that they are safe for human consumption.

A healthy environment, shelter and food are things we have to make and produce. We need to grow food in our gardens. We need to build shelters, whether a permanent house or a traditional bush house. We need to make flowerbeds and cut lawns. We build footpaths and fences around our houses to create a safe healthy environment.

For you to try

- Go on an excursion or walk around your local community and keep the environment clean by cleaning up rubbish and polluted areas – such as school grounds, local markets and streams.

2. Opportunities

Papua New Guinea is a developing country. Young children need to go to school to get formal education. They need to gain skills and knowledge, not only for higher education and employment but also for better living.

Young people need to learn to:

- make and produce things to use and sell and so become successful in business
- build good relationships within their own community and with people of different cultures and languages in Papua New Guinea
- build good relationships with other countries in the world
- develop skills and talents in sports

For you to try

- Identify an item that can easily be made or grown in the classroom or the school grounds, e.g. artifacts, growing food.
- Make the item or grow the food and then take it to the market and sell.

3. Importance of teamwork PD

Teamwork is important so that our school community, local community and nation can grow and advance. We need to:

- share responsibilities
- work as a group with members of our families and within the school and local community

The major principles of teamwork include:

- sharing ideas and learning from each other
- sharing responsibilities and co-operating with each other
- listening to each others' problems and complaints to reach a shared understanding
- group discussions that lead to democratic decision-making

For you to try

- Initiate a school beautification project, such as planting trees, clearing rubbish, making garden beds, fixing up broken furniture.
- Plan and carry out the project.

4. The school environment: needs and opportunities

Clean buildings and beautiful surroundings are something to be proud of. All teachers and students should strive for a healthy and attractive school environment. Whether your school is rural or urban, all schools have problems and needs:

- **physical needs** such as maintenance to buildings, furniture and tools
- **environmental needs** such as keeping school grounds, lawns, paths, flowerbeds and farms clean and healthy
- **economic needs** such as school fees. Subsidies from the government are usually not sufficient to run the school

Everyone in the school community should address these needs because without a group effort, needs will not be met.

Here are some ways the school community (teachers, parents and students) can address its needs:

- Organise fund-raising activities in the school that involve parents and the local community so that economic needs are met.
- Arrange working days for the local community to help with environmental problems relating to school buildings and grounds.
- Write letters to local organisations and businesses to fund specific projects, for example, the purchase of library resources.

For you to try

- Survey all upper primary students to discover the most pressing needs of your school. Present this information under the following headings:
  Physical needs
  Environmental needs
  Economic needs
- Prioritise these needs and identify the most important need.
- Write a report of your findings and present it to your school principal.

One way to record your data

The School Environment – Needs			
Name of school:			
Needs	**Areas (three most needed projects)**	**Agree**	**Disagree**
Physical needs	1. 2. 3.		
Environmental needs	1. 2. 3.		
Economic needs	1. 2. 3.		

SCHOOL IMPROVEMENT PROJECTS

1. Planning strategies and procedures

When starting out on any project, the following steps should be taken.

1. **Investigate.** Ask questions to find out the specific needs of the school community and how these can be met. This information can be collected by:

- writing a questionnaire and canvassing oral opinions from students
- conducting a survey
- listing five possible projects and asking students to vote to determine which project is top priority

2. **Plan and design.** Once the project has been determined, you need a plan of action. Students need to be organised into groups. Give each group a specific objective. At this stage students must be able to:

- design the object (e.g. fence for school garden) or school improvement project (e.g. repairing school desks) and this includes writing a list of resources needed to make or complete the project
- write the procedures for completing the project as well as a proposed time-line for completion

3. **Implement the plan.** Work co-operatively with the team to complete the task. Appoint a team leader who will:

- give every member of the team a task
- make sure the project is progressing to plan
- manage time and resources effectively
- address problems if they should occur

4. **Evaluate.** Once the project has been completed it is important to evaluate the project.

- How clear were the plans and procedures?
- Did the students work as a team?
- How reasonable was the schedule?
- Were there problems that made it difficult to complete the project?
- Can this project be extended or developed further?
- Write a report on the processes involved in the project.
- If money was involved, write a financial statement for the project.

2. School projects

Project 1. Repainting old chalkboards

Chalkboards in our classrooms are an important way for teachers to convey written information. Students read information from chalkboards to undertake set tasks. Continuous writing and rubbing causes the chalkboard paint to fade and this makes writing difficult to read. Maintaining chalkboards is a useful project as all classrooms must have a well-functioning chalkboard.

Tools and materials needed

Chalk duster and piece of cloth
Bucket of water and cleaning detergent
Paint brush
Chalkboard paints (green or black in colour) from hardware stores
Paint cleaning detergent – turpentine

Procedure

1. Check all classrooms and identify chalkboards that need repainting.
2. Estimate materials required and their cost.
3. Clean and wash the chalkboard thoroughly and leave it to completely dry.
4. Mix the paint carefully and thoroughly.
5. Using the brush, apply the paint evenly onto the chalkboard surface. Start from one end and work to the other; work from top to bottom.
6. Leave the chalkboard to dry in the shade, free from dust (24 hours drying time).
7. Try writing on the newly painted board to see if the writing is legible from the back of the classroom.
8. If still illegible, re-apply the paint but this time very lightly, and leave it to dry.

Project 2. Maintaining old desks

Schools do not have extra money to buy new desks so a project to repair and maintain desks saves money and makes the classroom a more attractive environment.

Many of the desks in the classrooms are very old and the following repairs may be needed:

- The desktop surfaces may be dirty, covered in writing or damaged. This makes it difficult for students to use the desk surface for writing.
- The joints to the seat and legs may be loose so that the desk wobbles.
- The metal desk frames may be rusty and need to be cleaned of rust.

Tools needed

Saws (crosscut saw, tenon saw)
Hammer (claw hammer)
Spanners
Screw drivers
Tape measure
Sanding block
Wire brush
Paint brushes

Materials needed

Plywood (type and size similar to old desktop)
Timber (type and size similar to old desktop)
Nails
Bolts and nuts (to match type used on the old desk)
Screws
Wood glue
Wood filler (putty)
Glass papers (sanding papers)
Paints (varnish)
Turpentine (kerosene)

Procedures

1. Check and collect all the damaged desks that need fixing.
2. Estimate the cost of the suitable materials that are required.
3. Collect all materials and tools needed.
4. Dismantle the damaged desk parts by pulling out nails and loosening bolts and nuts.
5. After removing the damaged parts of the desks, take measurements and transfer these to a new piece of plywood or timber.
6. Cut the new pieces with a saw.
7. Assemble the desk parts by nailing, screwing and using bolts and nuts.
8. For metal frame desks, brush off the rust using a wire brush.
9. Paint the metal desk frame using a metal-coating paint and leave it to dry.
10. Sandpaper the desk tops, seats and legs thoroughly.
11. Wipe out the dust.
12. Stir the paint well and apply the paint to the desk using a paint brush.
13. Leave them to dry in a place free from dust and out of direct sunlight.

Project 3. Church notice board

The church notice board is used to give information to members of the community about such things as service times, church programs, community concerns, comments and projects.

Materials needed

2 posts – bush timber – 20-30 cm in diameter, height 230 cm (iron posts could be used)
2 cross-bar timbers – 20-30 cm diameter, length 240 cm
1 piece particle board/plywood – 2400 mm × 1200 mm × 18 mm
Some pieces of bush timber – length 150 cm for rafters
Bamboo splits or small bush timbers – length 340 cm for cross rafters
Bundles of palm leaves or kunai grass for the roof
Nails – 6 inch, 3 inch and 2 inch nails, flathead
Bush twines/robes for fastening palm leaves/kunai grass to the rafters

Tools needed
Bow saw
Tape measure
Bush knife
Crosscut saw
Claw hammer

Procedure

1. Make a working drawing.
2. Estimate the cost of materials needed.
3. Collect the materials and tools needed.
4. Select and mark out the working site.
5. Measure the posts and cut to required length.
6. Dig holes, erect and secure the posts.
7. Secure posts together by nailing the crossbar and secure timbers to the top end and middle of the posts.
8. Fasten the roof joist onto the posts.
9. Secure the rafters onto the roof joist.
10. Fasten the bamboo splits onto the rafters.
11. Firmly tie the palm leaves or kunai grass on to the roof.
12. Secure the particle board onto the posts.
13. Nail strips around the board edges.
14. Lay sand and gravel around the base of the notice board.

Project 4. Drum oven

Drum ovens are made from empty fuel drums and are quite popular in Papua New Guinea. Drum ovens are good for baking bread (or cakes or scones) in villages that have no electric or gas stove ovens.

There are many different ways to make a drum oven. Here is one of the ways to make a drum oven. Note, the instructions are brief. At all times use common sense and include more steps if you think they are necessary to successfully complete your drum.

Tools needed
Hammer
Spade
Bucket
Trowel
Hack saw
Cold chisel
Tape measure
Bolt cutter

Materials needed

One 200 L drum – must be in good condition
Galvanised pipe 1.5 m long
Measured and cut arc-mesh wire that correctly fits into the drum
Fine gravel, cement, stones, water, chicken wire

Procedure

Preparing the drum

1. Using the cold chisel and hammer, chisel out the top lid.
2. If there is still oil on the inside of the drum, make a fire inside the drum to burn it off.
3. When the drum is cold, wash it out with water.

Building the oven

1. Position four large stones to form the base and place the drum on its side on top of the rocks. The space under the stones become the fire place
2. Put stones at the back end of the drum and stand the pipe for the chimney on top of the drum at the back.
3. Wrap around the drum with chicken wire to hold the mixed cement and stones from sliding or falling off the drum.
4. Mix the cement and fine gravel together with water.
5. Pile the stones up together by placing the concrete mixture in between the stones.
6. Make sure to stand the pipe for the chimney up straight when piling the stones up.
7. Cover the two sides and the back completely with stones and concrete mixture.
8. Cut a piece of arc mesh for the shelf inside the middle of the oven. The food to be baked is put on this shelf and is cooked by dry heat.
9. Use the chiselled out top lid of the drum as the door of the oven. Nail a small piece of wood to the outside centre of the lid as a handle.

Additional small school projects

1. Line river stones along the flowerbeds.
2. Line old used tyres along the edges of paths.
3. Cut up old tyres and use them as pot plants.
4. Make stick or pitpit fences around flowerbeds to prevent people walking over them and pigs from digging up the flowerbeds.
5. Make a flower garden around the school flagpole.
6. Make classroom decorations and displays.
7. Use old fuel drums as school or rubbish drums.
8. Involve the whole school in a school beautification project.
9. Involve the whole school in cleaning up a disaster-stricken areas (for example, after cyclone, landslide or flood damage).
10. Involve the whole school in road maintenance.
11. Involve students in local community activities like maintaining the village water supply, road maintenance or the village clean up.
12. Encourage students to carry out awareness programs on important issues.
13. Involve students in fund-raising for community projects or school projects.

Additional Material: Project Sheets

Project Sheet 1.

Shoe mud wiper/remover

Use

The mud wiper is used to wipe out mud or sticking soil under your shoes or feet. This small simple project helps to prevent you taking mud into your house, classroom and offices. The advantage of this project is that it is very light to carry around and easy to make from off-cut pieces of timber and metal. You do not need to buy materials to make it.

These can be made and sold in the market to generate some income.

Materials needed

2 pieces of '6 × 1' timber, 20 cm in length
30 cm long metal (broken grass knife or bush knife cutting blade)

Tools needed

Crosscut saw
Hacksaw
Hammer

Steps

1. Collect pieces of timber as stance and metal.
2. Measure and cut two pieces of timber ('6 × 1') to 20 cm in length, and metal to 30 cm or to desirable length.
3. Measure and cut a deep straight cut in the middle of the top ends of the stances. Measure 4 cm in from the sides for the bottom end and make a 'V' cut to ensure good balance of the stances. Measure 4 cm from the sides of the top end of the stance and cut off the sides.
4. Insert the thin metal blade into the middle deep cut – the size to that of the metal width.
5. If the blade is inserted loosely, nail the metal and timber firmly.
6. Finish off by sanding smooth and apply varnish.
7. Leave to dry completely.

Project Sheet 2.

Coconut brooms

Use

Coconut brooms are used to sweep rubbish inside and outside homes in Papua New Guinea. Broom sticks are made from coconut leaves or from oil palm leaves.

Materials needed

Matured long coconut leaves or oil palm leaves
Strong bush vine for rope

Tools needed

Bush knife
Small kitchen knife

Steps

1. Climb coconut trees and cut matured leaves. For oil palm tree leaves, use a sickle to cut the leaves down.
2. Cut off the leaves from the stem.
3. Remove the leaves off the coconut broom sticks.
4. Bundle the broom sticks together (handful size) and tie with vine rope.

Project Sheet 3.

Stuffed man or meri doll

Use

The doll is used as a safe toy for babies or small children. It can be clothed to represent a male or female.

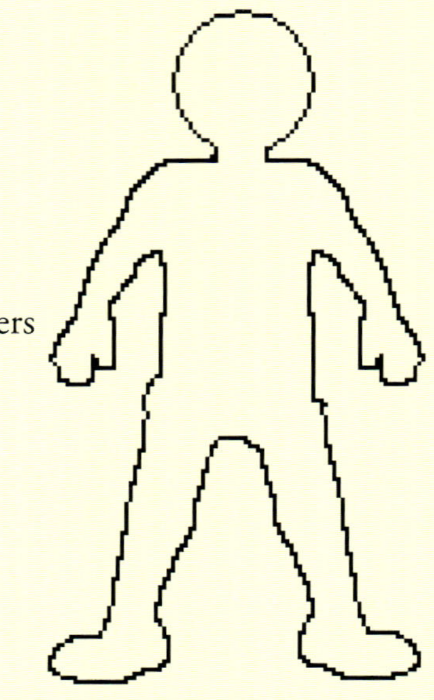

Materials needed

Old paper
Brown scrap fabric
Coloured scrap fabric
Black wool
Black, white and red fabric paint or markers
String to make a bilum or arm band
Kapok or cotton wool for stuffing

Equipment needed

Sewing equipment: needles, thread, pins, scissors, sewing machine (if available)
Thin stick for pushing stuffing in

Steps

1. Make a paper pattern. Draw the outline of a body on a piece of paper. This is the line for stitching. Then 5 mm from this line, draw another line to represent the seam allowance.
2. Pin the pattern onto a double piece of brown fabric. Cut carefully around the pattern edge. Remove the pattern.
3. Either by hand-stitching or by machine, sew strong, small stitches 5 mm from the outer edge to join the two pieces of fabric together, allowing an opening at the top of the head.
4. Turn the fabric inside out, using a thin stick to get the shape right, without breaking the stitching.
5. Paint or draw the eyes, eyebrows, nose and lips on the face.
6. Stuff the doll shape with soft stuffing.
7. Wind the black wool around a narrow length of card about 8 cm long to make the curly hair. Stitch the wool onto the top of the head and remove the card.
8. For a meri doll, use scrap fabric to make a meri blouse and laplap. Use string to make a small bilum to hang on the doll's shoulder.
9. For a man doll, use scrap fabric to make a laplap. String or vine can be used to make an armband.

Note: Other soft toys can be made in similar ways using shapes such as a fish or pig. Fins or ears can be attached and face features sewn using embroidery stitches.

Project Sheet 4.

Sun shade hat

Use

The hat is used to shade the face.

Materials needed

Old plastic container or other firm material for shade
Vine to weave the head band
Shells or seeds for decoration

Equipment needed

Tape measure
Cutting knife

Steps

1. Make a paper pattern. Draw a semi-circular shape for the hat-shade on paper and adjust the measurement and shape to suit a person's head.
2. Place the pattern on a piece of plastic bucket or ice cream container or other suitable material. Allow 2 cm on the head edge to attach to the head band.
3. Carefully use a sharp knife to cut the hat-shade shape.
4. Measure the distance around a person's head. Using suitable vine or bush rope, weave a head band. A design can be woven into the head band.
5. Be creative in devising ways to attach the hat-shade and decorative shells or seeds to the hat band.

Note: Be creative and innovative and design other hat styles using bush materials or scrap fabric.

Project sheet 5.

Orchid nursery

Use
The orchid plants can be grown to beautify the school grounds and to sell in the market.

Materials needed
10 strong and short logs about 2 m long of any size
A roll of tie wire about 20 m long
A small shelter (4 m × 4 m) and 2 m in height (nursery)

Equipment needed
10 different variety of orchid plants
Spades for digging holes
A green nursery cloth or green fly wire

Steps
1. Select good flat land near the school to build the nursery.
2. Mark the holes for the orchid posts with equal distances and height (five on each side).
3. Carefully place an orchid plant on each post perfectly upright and tie with a wire.
4. Support the orchid plants with coconut husks or banana fibres around the base and tie with wire firmly.
5. Build a strong fence around the nursery.
6. Water your plants in the afternoon and watch out for pests.
7. Plants should produce flowers within six to ten months.
8. Harvest the flowers and cuttings with a sharp knife and bring them to the market for sale.

Note: Repair the nursery with strong timber from the bush.

Project sheet 6.

Planning your vegetable plot

Vegetable planting chart

Instruction	Tomatoes	Cabbage	Beans	Carrots	Corns
Size of seed	Medium	Medium	Large	Small	Large
Depth of sowing	1 cm	1 cm	4 cm	1 cm	1.5 cm
Sow in seed bed	Yes	Yes	No	Yes	No
Row spacing in seed bed	15 cm	15 cm	No	No	15 cm
Thin in seed bed to	2 cm	3 cm	No	No	No
Sow direct in plot	No	No	Yes	No	Yes
Row space in plot	60 cm	45 cm	20 cm	30 cm	60 cm
Space between plants in row	50 cm	45 cm	5 cm	1 cm	50 cm
Thin in plot to	No	No	10 cm	2 cm	No
Days from planting to harvest	100	80	50	90	90

Use

Planning your school garden is important because it will help you to plan and grow your local vegetables according to the amount of land available.

Materials needed

Good land for farming near the school
Variety of local vegetable seeds
Nursery
Garden tools

Equipment needed

Writing pad
Pencil
Ruler
String-line
Tape measure

Steps

1. Draw a plan of your vegetable plot. (This helps you to get the spacing of the rows right.)
2. In planning think about the height of the plant and the amount of room it needs when fully grown. If plants are too close, they will crowd each other. They will compete for light, water and nutrients. If they are too far apart, they will waste the space on the plot.
3. Look at the vegetable sowing chart on this page and use it to help you draw a plan in your exercise book.
4. First decide which vegetables you are going to grow, then work out how much space they each need.

Note: Be creative and innovative and design plots for other local vegetable groups that grow well around your school.

Additional Material: Projects and Investigations

MANAGING RESOURCES

1. Vegetable garden

Select an area in your school for a garden. Select local vegetables like aibika, chinese cabbage, round cabbage, cucumber, tomato, pumpkin, beans, egg-plant and capsicum. Sell your produce at the local market or to teachers at the school.

2. Raising chickens

Build a small kunai or sago-roof house with one room. Ask parents to donate local chickens to the school for their children's agriculture lesson. The birds should be looked after and housed properly. Feed them with local food. Sell eggs to staff and students.

3. Raising pigs

Build a small kunai or sago-roof house with two to three strong rooms. Ask parents to donate local pigs to the school for their children's agriculture lesson. The animals should be cared for and housed properly. Feed them with local food and sell the pigs at the age of nine to ten months. Pigs should be sold before the end of the school academic year.

4. Building a fish pond

Select a running creek near your school and change it into a fish pond. Make a fence around it and put in some fresh water fish. The best local fresh water fish would be: tilapia, golden carp and cold fish. Sell the fish to your teachers and parents.

5. Orchid nursery

Select an empty space in your school for an orchid garden. Collect local orchids from the village and plant them in your garden. When established, take cuttings and sell them to local people or visitors to your school.

6. Shellfish farming

Select a shallow sea water area near a reef or in a lagoon. Fence an area measuring 4 × 4 metres using mangrove poles. Tie the poles with wire or cane ropes. Next place your shellfish inside. Harvest them after six months and sell them at the local market in your area.

7. Cut flower arrangements and flower leis

Consider planning a project to sell cut flower arrangements or flower leis. Create gardens of flowering plants. Plant flowering trees in your school grounds too.

8. Compost

Set up an area for a compost heap. Collect organic matter. Use the compost to enrich the gardens.

BETTER LIVING

1. Belts, necklaces, arm and ankle bands

Gather resources and learn the skills to make belts, necklaces, arm and ankle bands to sell. Make imaginative use of local materials for a truly Papua New Guinean look.

2. Brooms and rakes

Investigate the need for new brooms and rakes in your community. If there is a need, make strong, attractive brooms and rakes from available resources. Determine an affordable price.

3. Clothes repair service

Plan a clothes repair service for children or adults. Consider charges according to the size of damage and materials and time required.

4. Gardening service

Establish a service to weed gardens, rake up leaves, trim hedges and beautify flower gardens. Determine an hourly rate per person and seek customers.

5. Hats

Plan a range of designs for hats for both males and females. They can be made from fabric or from local materials.

6. Kitchen utensils

Consider which kitchen utensils could be made from local or recycled materials. Items could be graters, spoons, knives, chopping boards, soap holders, pot stands or tongs. Make them simple, useful and attractive.

7. Learn to read and write

Explore whether there are people who would like to learn to read and write. Cater for individual or group interests. Calculate a reasonable cost for each hour of teaching.

8. Plant cuttings

Investigate whether people would be prepared to buy plant cuttings. Take cuttings from plants in the garden. Sell in bundles for one or two kina a bundle.

9. Reading materials

Collect old newspapers, magazines or books. Look in second-hand shops for books that could be re-sold. Advertise reading materials for sale at reasonable prices.

10. Removal service

Offer to provide quotes for removal services – no job too big or too small. It could be rubbish or furniture removal. Calculate a reasonable cost according to the number of people and the time needed to do a job.

11. School lunches

Plan a variety of foods that could be sold to children on different days. You could sell ice blocks, fruit drinks, hard-boiled eggs, sausages on sticks, sandwiches, doughnuts, peanut toffee, chips, soup or fruit. Sell at affordable prices.

12. Special food for special events

Develop skills to make iced cakes or attractive finger-food for special celebrations. Let others know that you can provide special food for special events.

13. Vases

Design a range of attractive vases or plant holders made from different local materials such as bamboo, coconut shell, clay, recycled plastic or glass containers, tyres or bush timber. Sell at markets, open days or parent gatherings.

COMMUNITY DEVELOPMENT

1. Community resource map

Draw a resource map of the community you live in. Then based on the resources map, plan and carry out a project to promote your communities resources. Suggestions: sell tapioca cake, establish a traditional craft shop or vanilla farm and so on.

2. Community water supply project

Your school community needs water tanks to supply clean drinking water to both teachers and students. Write a letter to your local ward councillor requesting financial help to establish two bushman water tanks in the school.

3. Drying fish – smoking method

Plan and build a small smoke-house in your community to dry fish.

4. Community Awareness campaign

Form small groups in your class and prepare charts, role-plays, write speeches, songs and poems to educate your community on current issues like HIV/AIDS, drug abuse, malnutrition and the importance of a clean living environment.

5. Classroom maintenance and repair

Your school has a lot of dirty and broken tables, chairs and bookshelves. They needed repair. How can you fix them?

6. Maintaining rusting roofing iron

Your classroom's roofing irons are rusting away. You need to prevent rusting. What can you do? Make a list of tools and materials required and write out a plan of action.

7. Collecting good, clean rain water

It usually rains heavily and you wish to collect rain water. But you do not have a gutter connected to your roofing iron for rain water to run through. What other bush materials could you use instead of a gutter and water tank? How would you collect rain water?

8. Building a small bridge over a creek

You have a small creek that runs through your village community or school. Sketch a plan for a bridge. Collect all materials needed to build the bridge over the creek.

9. School sign board

Your shool does not have a sign board. Draw up a plan for your sign board and collect all the materials and tools needed. Build a school sign board.

10. Design a community game

- Make a board game with 40 squares – each square numbered from 1 (start) to 40 (finish).
- Write five messages that benefit a community.
  Example: PMV bus service is always on time
- Write five messages that don't benefit the community.
  Example: Rubbish is dumped in the river
- Write your ten messages on random squares.
- Write instructions on how to play the game, then play the game with your friend.

Making a Living Game

RULES

- Each player in turn throws a dice and moves the designated number of spaces.
- If you land on a square with text, read the statement and explain why you deserve to go forward or back. Note: the group has to accept your explanation as sensible before you can play on.

Crop rotation used. Move forward four spaces.
8
9
10
11
Frost. Miss a turn.
12
13
14
15
16
17
18
19
20
Compost used in garden.
21
Go forward six spaces.
22
23
24
Landslide – road blocked. Go back four spaces.
25
steal
Go
the
no
a turn.
43
44
Ground well mulched.
45
Move forward two spaces.
46
47
48
49
50
51
Fresh fish left out in the sun. Go back three spaces.
52
53
54
55
livestock
Have
turn.
in chicken
Miss a turn.
65
FINISH

Glossary

abundance	plenty, more than enough
adapt	to change and become suited to something new
aesthetic	artistic, beautiful
aibika	a common green vegetable in Papua New Guinea, good for boiling with chicken or pork to make soup
bake	to cook food in an oven, or an enclosed space where the heat is dry
balanced meal	a meal that is good for you because it has some food from each of the food groups
bank statement	a record from the bank showing details of money in an account
bilas	decoration
biodiversity	the variety of plants and animals in an environment
boil	to cook food in very hot bubbling liquid
broadcast	a radio or television program
budget	a plan to spend money
carbohydrate	the main source of energy in food, found in sugar and starch, for example
care	to look after the well-being of someone or something
clam	a large shellfish, with white and grey flesh
communication	passing news or messages from one person to another
community	a group of people living in the same locality who share something in common
conserve	to use something wisely so it will not run out, or be damaged
consumer	someone who buys or uses something
consumption	eating or using up something
contour	a line on a map which joins points that are at the same height above sea level
cover crop	a crop in which other plants or animals live

criteria	the standard by which you judge something
crotalaria	a legume plant – otherwise known as *botol plauwa* in Tok Pisin
culture	the customs, traditions and beliefs that make up our way of life
deforestation	cutting down trees in a forest. The opposite of reforestation
democratic	a belief that everyone has equal rights
disease	illness or sickness
disposal	throwing something away
dispute	a disagreement or argument
element	an important part of something
enterprise	a type of business
environment	the world around us, and the conditions that affect growth and development
equipment	the things you need to do something
essential	the most important things, which we cannot do without
evaluate	to decide the value of something
expenditure	spending or payments
exploration	discovery
export	to sell something overseas, like gas, gold or coffee, to earn money for our country
fats	greasy or oily nutrients in food which are a source of energy
finance	to do with money
flammable	something that can easily go on fire
fry	to cook something in a pan with oil or fat
furniture	tables, chairs, cupboards, beds and other moveable things we have in our houses or schools
galip	a type of nut, or the tree it comes from
generate	to create something or make something happen
geology	the study of the Earth and rocks
grill	to cook food by heat given off from very hot coals or a hot cooker
healthy	in good condition, free from sickness
heritage	the values and customs passed down to us from our ancestors
hygiene	cleanliness
income	money that you are paid for doing a job
influence	to have an effect on someone or something

interdependent	when things depend on each other
interview	to ask questions and collect information
investigate	explore, examine, look into
invoice	a bill for money that is to be paid to someone for something
kumu	a name for green leaves that you can eat
labour	the work someone does
laulau	a soft fruit that you can eat
laundry	the job of washing and ironing clothes
leucaena	a legume plant – otherwise known as *marmar* in Tok Pisin
lifestyle	the way people live their lives
making a living	the things we do to earn money to help us live well
mammals	humans and animals who feed their young with their mothers' milk
management	a planned way to do something
mangrove	a tree that grows in salt water
market	a place where things are bought and sold
media	newspapers, radio, television
minerals	things like iron and calcium, found in food, which help keep us healthy
modern	whatever is happening today, the latest ideas or fashions
modify	to change something
nutrition	food that helps keep us healthy
pau	a type of nut that hangs in bunches
pests	destructive or annoying insects or animals, like cockroaches or rats
plan	a way to do something, like a diagram
procreria	a legume plant
produce	make, create
productivity	the way in which a lot of things can be produced
project	a plan to make something
propagate	to reproduce new plants using parts of existing plants
protein	things in food that help us to grow well
puereria	a legume plant – otherwise known as *rop plauwa* in Tok Pisin
pylons	the towers that support electric power lines
pyrethrum	a plant grown, especially in the highlands, to make pesticide

questionnaire	a list of questions to gather information
receipt	a written proof of a payment that has been made
recycle	to reuse something well, instead of throwing it away
resistant	hard to break, hard to get through
resources	anything that can be used to create wealth, or make life better
revenue	income or money that comes into our country, or to our school, or village
rotation	changing the crops that are planted in a garden each planting season
rural	to do with villages or country areas, not towns or cities
safety	the way in which we protect someone or something from danger
sediment	the soil that is carried away by a river, especially in a flood
staple food	the main food eaten by people, e.g. sago, sweet potato, taro, yams, rice
sustainable	to support something and keep it going
tools	things we use to get something done, or fix something
traditional	ways of the past, old ways of doing things that have not changed for hundreds of years
unemployed	not working in a paid job
unenclosed	not having a fence or wall around it
urban	to do with the towns, not the villages or rural areas
vitamins	important things in food that keep us healthy
yield	the amount of produce harvested

Acknowledgments

The authors and publishers wish to thank the following authors and copyright holders for granting permission to reproduce their material:

Crawford House Publishing Australia: pp. 2 (middle), 6, 10 (right), 12, 13, 14, 16, 18, 19 (left & right), 20, 23, 33 (top & bottom), 37, 44, 46 (bottom left), 70 (left), 76 (middle, bottom left), 77 (top & bottom), 80, 81, 82, 84 (top), 85 (left), 87 (top and bottom), 88, 90, 91; Barbara Hodgins: p. 46 (middle); Colette Modagai: p. 42; Pam Norman: pp. 110, 111; PhotoDisc: pp. 9, 96, 99, 116; Stephen Ranck: pp. 2 (top), 4, 8, 10 (left), 11, 22 (all), 26, 31, 40, 43, 45, 47, 55 (left and right), 70 (right), 75, 76 (bottom right), 83, 84 (middle & bottom), 85 (right), 89; Irene Sawczak: pp. 2, (bottom left and right), 5, 25, 28, 29, 32, 38, 46 (top and bottom right), 49, 52, 53, 60, 61, 66, 76 (top), 78, 86, 97, 98, 100, 101, 103, 106, 109.

Cover: (top) Irene Sawczak; (middle) Local Kavieng fishermen with net © Michael Gebicki, LONELY PLANET IMAGES; (bottom) Crawford House Publishing Australia; (far right) Irene Sawczak.

Every effort has been made to trace the original source of copyright material contained in this book. The publisher would be pleased to hear from copyright holders to rectify any errors or omissions.